SEMICONDUCTOR CERAMICS
Grain Boundary Effects

ELLIS HORWOOD SERIES IN PHYSICS AND ITS APPLICATIONS

Series Editors: Professor MALCOLM J. COOPER, Department of Physics, University of Warwick; and JOHN W. MASON B.Sc., Ph.D., Scientific and Technical Consultant

Dobrzynski, L., Blinowski, K. and Cooper, M.   NEUTRONS AND SOLID STATE PHYSICS
Dyson, N.A.   RADIATION PHYSICS WITH APPLICATIONS IN MEDICINE AND BIOLOGY: Second Edition
Elwell, D.   PHYSICS FOR ENGINEERS AND SCIENTISTS
Gough, W., Richards, J.P.G. and Williams, R.P.   VIBRATIONS AND WAVES
Granier, R. and Gambini, D.-J.   APPLIED RADIATION BIOLOGY AND PROTECTION
Hasnain, S.S. (editor)   SYNCHROTRON RADIATION AND BIOPHYSICS
Hozer, L.   SEMICONDUCTOR CERAMICS: Grain Boundary Effects
Ignatowicz, I. and Kobendza, A.   SEMICONDUCTING THIN FILMS OF $A^{II}B^{VI}$ COMPOUNDS
Martin, J.L.   GENERAL RELATIVITY: A Guide to its Consequences for Gravity and Cosmology
Rosse, W.G.V.   AN INTRODUCTION TO STATISTICAL PHYSICS
Rowlands, G.   NON-LINEAR PHENOMENA IN SCIENCE AND ENGINEERING
Saleem, M. and Rafique, M.   SPECIAL RELATIVITY: Applications to Particle Physics and the Classical Theory of Fields
Scott, V.D. and Love, G. (editors)   QUANTITATIVE ELECTRON-PROBE MICRO-ANALYSIS 2nd Edition
Steward, E.G.   FOURIER OPTICS: An Introduction, Second Edition
Trevena, D.H.   STATISTICAL MECHANICS: An Introduction
Wadas, R.   BIOMAGNETISM
Whorlow, R.W.   RHEOLOGICAL TECHNIQUES, Second Edition

SEMICONDUCTOR CERAMICS
Grain Boundary Effects

LESZEK HOZER

Electronic Materials Research
and Production Centre, CeMat'70, Warsaw, Poland

Translation Editor

DIANE HOLLAND

ELLIS HORWOOD
NEW YORK LONDON TORONTO SYDNEY TOKYO SINGAPORE

PWN — POLISH SCIENTIFIC PUBLISHERS
WARSAW

English edition first published in 1994
in coedition between
ELLIS HORWOOD LIMITED
Market Cross House, Cooper Street,
Chichester, West Sussex, PO19 1EB, England
A division
of Simon & Schuster International Group
and
POLISH SCIENTIFIC PUBLISHERS PWN Ltd.
Warsaw, Poland

Translated from the Polish by Jolanta Krauze

COPYRIGHT NOTICE:
© Polish Scientific Publishers PWN Ltd., Warszawa 1994

All Rights Reserved. No part of this publication may be reproduced, stored in a retrieval system, or transmitted, in any form or by any means, electronic, mechanical, photocopying, recording or otherwise, without the permission of Polish Scientific Publishers

Printed in Poland by D.N.T.

ISBN 83-01-11306-5 (PWN, Poland)

British Library Cataloguing in Publication Data

A catalogue record for this book is available from the British Library

ISBN 0-13-808049-6

Library of Congress Cataloging-in-Publication Data

Available from the Publisher

Table of Contents

Chapter 1 — Introduction 1

Chapter 2 — Controlling the properties of grain boundaries during the fabrication process 3
   2.1 Technological processes 3
      2.1.1 Preparation of the powders 4
      2.1.2 Initial densification 8
      2.1.3 Sintering 10
   2.2 Energy band structure of grain boundaries 15
      2.2.1 Formation of a potential barrier on the surface of a semiconductor 15
          2.2.1.1 Effect of an external electric field 16
          2.2.1.2 Contact potential barrier 17
          2.2.1.3 Surface states 19
      2.2.2 Shape of the potential barrier 21
      2.2.3 Flow of a current through the potential barrier 22
          2.2.3.1 Thermionic emission 23
          2.2.3.2 Schottky emission 24
          2.2.3.3 Tunnelling 26
          2.2.3.4 Other effects 28
   2.3 Segregation effects at the grain boundaries 29
   2.4 Adsorption and catalysis 32
   References 39

Chapter 3 — Metal-oxide varistors 44
   3.1 Properties of ZnO and Bi_2O_3 51
      3.1.1 Zinc oxide 51
      3.1.2 Bismuth oxide 53

3.2 Electrical properties of ZnO varistors 55
    3.2.1 Current-voltage characteristics 55
    3.2.2 Dielectric properties 59
    3.2.3 Parameters of the potential barriers 63
    3.2.4 Other properties 65
3.3 Microstructure of ZnO varistors 65
    3.3.1 Two-component systems 66
    3.3.2 Multicomponent systems containing Bi_2O_3 68
    3.3.3 Multicomponent systems with addition of rare earth metal oxides................................... 75
    3.3.4 Microstructure of grain boundaries................ 76
3.4 Mechanisms of electrical conduction in ZnO varistors 79
3.5 Degradation of the electrical properties of varistors 86
    3.5.1 Effect of degradation upon the J–V characteristic... 87
    3.5.2 Effect of degradation upon the dielectric properties of a varistor 88
    3.5.3 Effect of heating upon the behaviour of degraded varistors 89
    3.5.4 Other effects associated with varistor degradation ... 89
    3.5.5 Models of degradation mechanisms............... 92
References 98

Chapter 4 — Positive temperature coefficient of resistivity (PTCR) thermistors........................... 109
4.1 Properties of $BaTiO_3$ 112
4.2 Process engineering and microstructure of PTCR materials 114
4.3 Electrical properties of PTCR materials................. 120
4.4 Operating mechanism of PTCR materials 127
    4.4.1 Energy band structure of the grain boundaries above the Curie temperature........................ 127
    4.4.2 Energy band structure of the grain boundaries below the Curie temperature........................ 130
    4.4.3 Electrical conductivity of the interior of barium titanate grains 131
        4.4.3.1 Defect structure of $BaTiO_3$ doped with La .. 131
        4.4.3.2 The kinetics of defect diffusion 134
        4.4.3.3 $BaTiO_3$ doped with Sb ions 135
        4.4.3.4 Critical doping level 136
    4.4.4 Formation of potential barriers on the $BaTiO_3$ surface 137
    4.4.5 Potential barriers at the boundaries of $BaTiO_3$ grains 138
    4.4.6 Two-stage PTCR effect........................ 140

Table of Contents

4.4.7 Conduction mechanism above the Curie temperature	141
References	143
Chapter 5 — Boundary layer capacitors	148
References	157
Chapter 6 — Ceramic gas sensors	159
6.1 Technology, design and properties of SnO_2-based sensors	161
6.2 Operating mechanism	176
6.3 Selectivity	181
6.4 Other materials suitable for fabricating gas sensors	186
References	187
Chapter 7 — Overview	191
Author index	193
Subject index	198

To my Parents and Wife

CHAPTER 1

Introduction

The development of new materials that meet present highly specialized requirements has recently become the major factor which determines technological progress. Ceramic materials, a great variety of which are now available, are increasingly used. Ceramic technology has to draw more and more from physics and chemistry but it continues to be an art as well as a science. New ceramic materials are being developed by trial and error, with the theoretical interpretation of their behaviour being worked out afterwards.

The structure of the surface of solid bodies and the physics of surface phenomena, such as mass transport, electric charge transfer, adsorption and catalysis, have not yet been fully examined, at least as far as the behaviour of actual materials is considered. However, many ceramic materials, whose properties are largely controlled by surface behaviour, are already manufactured. This book is only concerned with certain representatives of this abundant group of materials.

The materials described in the present book include those designed for PTCR (Positive Temperature Coefficient of Resistivity) thermistors composed of $BaTiO_3$, related materials intended for GBBL (Grain Boundary Barrier Layer) capacitors and the materials for fabricating ZnO-based MOVs (Metal-Oxide Varistors). The properties of all these materials, which have been in use for a relatively long time, are now quite well known and described in the literature. The fourth group dealt with in the book comprises the materials intended for CSSDs (Chemically Sensitive Semiconductor Devices), i.e., gas sensors, which are mostly fabricated of SnO_2, ZnO and TiO_2. These materials are now in a phase of rapid development, but, as mentioned earlier, the theory remains far behind the engineering skills.

Certain chapters of the book, such as that concerned with the microstructure of varistor-type materials, present the evolution of opinion on the subject. We can see how difficult it is to examine certain phenomena even when modern measuring equipment is available. It also happens that the subjective view of an investigator may hinder the objective knowledge from dispersing more widely.

Section 2.2 gives a brief summary of the basic phenomena that occur during the manufacture of ceramic materials. Section 2.3 describes the fundamental properties of the potential barriers formed at the surface of a solid body. Here, the author has resorted to compromise between a detailed description and a brief outline, referring the interested reader to appropriate monographs.

The author believes that the book will help to break down the stereotyped thinking about ceramic materials, in which ceramics are commonly identified with ceramic art materials, constructional materials or, at best, refractory materials. The vast group of ceramic materials comprises very many non-metallic, inorganic materials which are manufactured using high-temperature technologies; even single-crystal semiconductors may be classified within this group. The book presents the extensive understanding about the phenomena that occur during the fabrication of ceramic materials and during the operation of devices made of them, but in order to complete the picture, abundant engineering data have also been included.

It is impossible in this short space to give an adequate statement of appreciation to all who have had a share in the preparation of this book. I wish to express particular thanks to Professor Andrzej Szymański for valuable discussions carried out during several years of our common work, and for his advice and optimism which help me to persist in continuous learning and working, first, on my doctoral thesis and, then, in writing the present book. I am also deeply indebted to Dr. Władysław Torbicz, Professor at the Institute of Biocybernetics and Biomedical Engineering of the Polish Academy of Sciences, for making much information presented in this book available and for his assistance in writing the chapter devoted to ceramic gas sensors. I am very grateful to the Kościuszko Foundation in New York for supporting my work at the Massachusetts Institute of Technology during the final stage of the preparation of this book.

CHAPTER 2

Controlling the Properties of Grain Boundaries During the Fabrication Process

Although, as already mentioned, the technique of fabrication of ceramic materials contains a large element of skill, it is increasingly being based on sound theoretical foundations. The present book aims at describing certain specific properties of selected ceramic materials, but it also indicates some fundamental understanding. Chapter 2 provides information about the process of fabrication of ceramic materials, during which the most important aims are to form an appropriate microstructure at the grain boundaries of the material and to ensure that they will have the desired properties, since these determine the properties of the final material (as far as the materials discussed in this book are concerned). Chapter 2 also describes briefly the phenomena that occur on the surface of a solid, such as the formation of potential barriers, carrier transport and certain phenomena associated with adsorption and catalysis. For more detailed description of these effects the reader is referred to the literature.

2.1 TECHNOLOGICAL PROCESSES

Modern ceramic materials are mostly fabricated of pure, chemically prepared, raw materials whose properties are precisely specified. Ceramic technology consists of initial densification and sintering of the powdered raw materials so that the phase and crystalline structures and the microstructure of the final product satisfy given requirements. The raw materials may be either simple compounds (e.g., oxides) of metals or more complex compounds that, during the fabrication process, decompose to form simple oxides. In order to ensure a desired phase composition of the powder mixture, the material is subjected to thermal treatments, such as calcination. Then, by milling, granulating and compacting operations, which are enhanced by the addition of suitable organic substances, we obtain the desired product in a rough-shaped form. The next operation is sintering during which the microstructure of the ceramic is finally established under the action of heat, ambient atmosphere and, in some cases, increased pressure. After the sintering process, we obtain a densified product with grains of desired average diameter and phase composition. In order to adjust its physical properties, the material is often subjected to additional heat treatment which modifies the structure of the boundaries between the grains, in particular their energy structure. The final stages of the fabrication of the component involve mechanical treatment, deposition of electrodes, attachment of leads and coating with protective plastic films.

The most important stages of the fabrication process will now be described in detail. It is of course important to know the mechanisms involved in the individual processes, since this permits optimization of the technology and the fabrication costs.

2.1.1 Preparation of the powders

The preparation of the starting ceramic powders has become almost the most important operation. The powder should have a specified chemical composition without unwanted impurities and its grains should be of the required size (usually as small as possible). It is often advantageous for the grain surface area to be highly developed, since the reactivity and the sintering capability of the powder is then increased. High surface area powders are prepared by disintegrating coarse-grained materials or using the particle growth techniques. The choice of an appropriate technique is made individually for a given material. The ideal technique would consist of disintegrating, to a submicron size, the grains of a powder whose composition, phase structure and crystalline structure are just as required. Such an ideal disintegrating technique would also need to be cheap and suitable for mass production, but has not however been devised yet.

The most common technique consists of milling the powders in *ball mills* (a rotating chamber filled with balls or cylinders made of a hard dense material of high specific weight). If, however, the grain diameter is smaller than 10 µm, this technique is ineffective since the powder easily re-agglomerates and the number of particles to be milled is too small. It is important that the mill balls have appropriate shapes and sizes since these affect the grain-size distribution of the powder. In order to avoid contaminating the ceramic with the ball material (that is, ground during the milling operation), the ball material should preferably have the same chemical composition as that of the ceramic material. This is, however, not always possible or effective. The rotating speed of the mill, its diameter and loading should be carefully selected. Mills that are not rotated but vibrated at a frequency of several tens of hertz appear to be more effective: the milling time is reduced by an order of magnitude thanks to the increased number of ball collisions. The milling process is most often conducted in a water suspension or, if required, in alcohol, acetone or the like. In order to make the process more effective, certain organic substances, such as acetamides, furfuryl acid, tiophene, furan, which prevent the agglomeration of ceramic particles, are often added. Vibrating ball mills are now used extensively for fabricating electronic ceramics on an industrial scale.

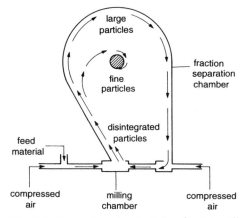

Fig. 2.1. Operational principle of a jet mill

Another milling techniques (using *jet mills* or *pulverizers*) utilize the collisions between the streamed powder particles passed through a decompressing gaseous carrier (air, steam, nitrogen), as shown in Fig. 2.1. The gaseous carrier, under a high pressure, decompresses in an appropriate chamber causing the ceramic powder particles to collide with one another and thereby to crumble. In certain mills, another counterdirectional stream of the gaseous carrier is passed through the chamber. This technique is very effective when the particles of the starting powder are small (< 10 μm in diameter), but some problems associated with the very fine final product may arise (the filters may become clogged).

Yet another technique uses that is known as an *attrition mill*, in which a suspension that contains both the material to be crushed and the milling substance are introduced in-between the walls of two cylinders (the internal wall of the stator and the external wall of the rotor which rotates at a high speed), where the particles of these two substances collide with one another. The as-milled powders are then separated mechanically or chemically. The efficiency of this technique is several times greater than that achieved with the vibrating mill, but the milling substance may heavily contaminate the ceramic powder.

The techniques increasingly used today for preparing ceramic powders are *particle-growth techniques* in which ceramic grains of desired composition, size and even shape are grown by, for example, chemical decomposition of a precursor (usually an organic-metallic compound).

Multicomponent ceramic materials are fabricated using *co-precipitation* from liquid solutions. The products, such as hydrated oxides, oxalates and the like, are subjected to an appropriate chemical treatment in order to give them the required final form. Since different compounds precipitate under different conditions, it is necessary that the solution should be stirred thoroughly so as to prevent the segregation of the constituents. A similar technique involves co-crystallization from complex salts that contain the required ions.

Many ceramic materials are fabricated by the *sol-gel method* in which sols of various substances are mixed together, gelled and then subjected to chemical and thermal treatments. This method is used, for example, for preparing powders with spherical grains. The co-precipitation and sol-gel methods may be very effective, provided that the process conditions are carefully selected for each material.

Another technique is *freeze drying* in which the pulverized drops of salt solutions that contain the required ions are abruptly cooled until 0.1–0.5 mm diameter spheres crystallize from them. Then, by moderate heating in vacuum, ice is made to sublimate, whilst the spheres containing the crystallized salts do not melt. The porous spheres thus obtained are oxidized and densified by calcination. In this way, by mixing mutually soluble salts, we can prepare many complex substances; if the cooling rate is sufficiently high, the segregation will be negligible. The grain diameter of the powders thus prepared may be of the order of nanometres. This technique is especially suitable for laboratory purposes.

Powders of very small grains can also be prepared using *plasma methods*, where, after an appropriate substance has been evaporated as a plasma, it crystallizes from the gaseous phase.

Most conventional technological techniques only permit preparation of one-component powders. In order to produce a material with a desired chemical composition, phase composition and crystallographic structure, these one-component powders are mixed in appropriate proportions and calcined, usually at a temperature much lower than the sintering temperature. Calcination can also be used for decomposing the starting materials, such as carbonates and oxides. The oxides obtained by calcination are often more reactive (thanks to the absence of surface contaminants and adsorbed gases) and can be sintered more easily. The conditions under which the calcination is carried out (the temperature, atmosphere, time) are selected for a given material according to the kinetics of the reaction in question (which may proceed in the solid or with the participation of a liquid phase). After the calcination, the material is powdered again, often several times.

The quality and properties of the as-calcined powders are examined using following methods:

(1) chemical and physical examinations, such as
— chemical analyses ('wet methods'),
— X-ray fluorescence spectrometry,
— mass spectrography,
— neutron activation analysis,
— atomic absorption spectroscopy,
— inductively coupled plasma emission spectroscopy (ICP),
— energy dispersion X-ray microanalysis (EDX),
— wave dispersion X-ray microanalysis (WDX),
— differential thermal analysis (DTA),
— thermogravimetry (TGA);

(2) crystallographic examinations:
— X-ray diffraction (XRD),
— electron diffraction,
— optical microscopy;

(3) powder morphology examinations:
— mesh analysis,
— sedimentation analysis,
— electrolyte conductance measuring method (the Coulter counter),
— optical microscopy with image analysis,
— scanning electron microscopy (SEM),
— measuring the specific surface area (e.g., by measuring the amount of adsorbed gas, known as the BET method),
— rheological examinations.

After the preliminary preparations described above, the ceramic powders are subjected to other treatments aimed at preparing them for initial densification. Additions are made of certain organic substances that facilitate granulation and enhance sliding during the pressing operation. Then the powders are granulated; this is done in mass production usually by means of a spray drier. Here, the powder suspension is mixed with organic substances in order to enhance the initial densification and to improve the homogeneity of the semi-finished product. The appropriate choice of these organic substances is difficult and each manufacturer usually uses his own recipes. For example, polyvinyl alcohol can be used as the binder, gum arabic as the deflocculant, Zn or Al stearate as the lubricating agent. Among other organic additives we may mention wetting agents (which improve the wettability of the surfaces), plasticizers, dispersants and substances preventing the ageing of the properties during storage. After mixing, the powder suspension is sprayed and dried in a stream of hot air. The individual powder grains agglomerate forming spherical granules. The spray drying technique is now most often used in industry since it is relatively cheap and easy to perform.

2.1.2 Initial densification

A variety of techniques may be used for preliminary densification of ceramic powders. We shall only describe those which are suitable for electronic ceramics. The most popular technique is what is known as *axial pressing*, exerted in a steel mould under a pressure between 10 and 200 MPa (Fig. 2.2). The figure shows an example of the pressure distribution throughout the material. We can see that the pressure and the rheological properties of the material must be matched in order that the product should be homogeneous. This can be done, and this technique may be used successfully in the mass production of small components, such as varistors and capacitors. The disadvantages of this technique lie in that only simple shapes can be pressed out, larger samples may be inhomogeneous (cf. Fig. 2.2), and the mould material (iron) may contaminate the sample surface. In the literature

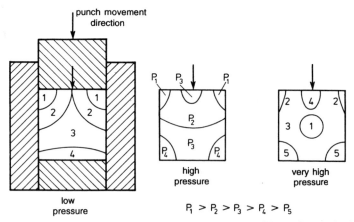

Fig. 2.2. Distribution of pressure in a ceramic compact subjected to uniaxial pressing in a cylindrical die (after Reed and Rink, quoted in Wang, 1976)

on the subject the reader will see how the properties of the ceramic granules, the kind and amounts of the organic additives and the pressing parameters (such as the friction against the mould walls) affect the efficiency of the operation.

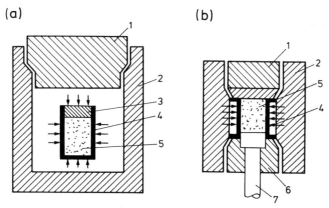

Fig. 2.3. Isostatic pressing: (a) wet method, (b) dry method; 1—upper punch, 2—chamber wall, 3—stopper, 4—elastic die, 5—ceramic powder, 6—lower punch, 7—sample ejector

Isostatic pressing, a much more modern technique, is based directly on Pascal's law. In its 'wet' version (Fig. 2.3a), an elastic chamber (made, e.g., of rubber or latex) containing the ceramic powder is acted upon by hydrostatic pressure (up to 1000 MPa) of water or of a special oil emulsion. With this technique, the densification of the product is much more effective and uniform than it is in the axial pressing method. Another modification of the isostatic pressure technique is 'dry' pressing (Fig. 2.3b), used especially in automated production, in which the elastic chamber is installed in a permanent manner in the apparatus and need not be removed from it after each pressing operation.

By choosing an appropriate shape of the elastic chamber, the isostatic pressing technique may be used for densifying components of sophisticated shapes, such as those with internal holes (if an appropriate rod is installed inside the chamber). The isostatic pressing technique is increasingly used for densifying ceramic powders; it is superior to e.g. casting, because of the much greater densities and better homogeneity of the final product.

Certain electronic components (such as capacitors) are fabricated using tape-casting, in which the ceramic powder with appropriate organic additives and a liquid medium (alcohol, water) in the form of a slurry is poured upon a moving metal (or plastic) belt. Excess fluid is removed with metal blades (doctor blade technique). In this way, we obtain a thin plastic ceramic tape, from which, after evaporation of the liquid carrier, we cut out (using steel cutters) components of required complex shapes. The tape technique is the basic method used for fabricating ceramic casings for integrated circuits.

The as-densified ceramic powder samples, having shapes close to those finally required, are subjected to sintering during which the properties of the material are eventually developed. We can, however, see from the above description that the preparatory operations are at least of the same importance as the sintering.

2.1.3 Sintering

During sintering, which involves the final densification of the ceramic material at a high temperature, most physical properties of the material are developed. The great amounts of thermal energy delivered to the system are used for

— growth of the grains and their densification,
— formation of appropriate chemical compounds,
— removal of the gaseous products of the chemical reactions,
— removal of the gaseous products generated by the decomposition of the organic substances added to the material (these are often removed by additional preliminary heating).

The sintering operation is carried out in many stages (heating up, annealing at a specified temperature, cooling), the atmosphere, temperature and duration of which should be carefully chosen. The material is often subjected to preheating at a lower temperature in order to produce what is called the 'biscuit'. The aim of this heating is to provoke chemical reactions in the solid state, to increase the mechanical strength of the material, and to prevent deformation and cracking of larger samples or samples of sophisticated shapes.

In terms of physics, the sintering processes are very complex and in different ceramic materials they can proceed in different ways. The literature only gives a few general descriptions of sintering mechanisms, such as the mechanisms and paths of the mass transport, the growth of grains and the behaviour of pores. Most investigators only report their experimental results, describing how the shrinkage and the density of the sintered material vary as functions of the temperature and duration of the process. The earliest models of the sintering mechanisms most often described are the one-component metal systems, where the mass transport was the easiest to examine. They assumed spherical grains, which was a considerable simplification. These models are not suitable for describing actual ceramic materials.

In order to facilitate constructing sintering models, it is necessary to divide ceramic materials into several classes that differ basically from one another. It is for example important to distinguish between single- and multiphase ceramic materials, since the participation of a liquid phase in the reactions, or the occurrence of certain reactions with the atmosphere plays a crucial role in the process.

During the sintering process, the system of loose powder grains tends to minimize its free energy. In single-phase and one-component materials, the excess of free energy depends on the extent of the grain surfaces and on the number of defects present in grains. In multiphase and multicomponent materials, the reactions occurring in the solid state and the presence of the liquid and gaseous phases cause mass transport processes to take place. During the sintering process, the centres of the grains come closer to one another, leading to an increase in the density of the material and a decrease in its porosity. When describing this process, we usually distinguish several stages, such as the removal of organic substances (1), densification (2), and the growth of grains (3), that differ from one another by the dominant densifying mechanism.

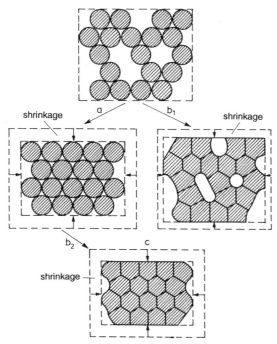

Fig. 2.4. Changes in grain configuration during sintering: a—regrouping, b—shrinkage (the grain centres come closer to one another), b_1—grain centres approach in a loose grain arrangement, b_2—grain centres approach in a close packed arrangement of grains, c—dense material (after Pampuch, 1988)

Pampuch (1988) describes how the configuration of the ceramic powder grains varies during the sintering process (Fig. 2.4.). Dereń et al. (1977) distinguish the following processes and mass transport mechanisms (movements of various elements of the structure) observed during sintering:

(1) movements of point defects:
— within the grain bulks (bulk diffusion, viscous creep due to diffusion),
— along the grain boundaries (diffusion along the grain boundaries),
— over free surfaces (surface diffusion);

(2) dislocation movements:
— sliding (plastic deformation of the grain as a whole),
— dislocation climb (creep due to this climb);

(3) mutual displacements of the grains (regrouping and changes of the grain arrangement);

(4) liquid phase transport — movements of atoms, molecules or other atom assemblies within the liquid phase (diffusion and viscous flow inside this phase);

(5) vapour phase transport — movements of atoms or molecules inside the gaseous phase (evaporation of the mass and its condensation somewhere else in the system).

Dereń et al. suggest that only three of these mechanisms can cause an assembly of grains to shrink and the porosity of the material to reduce. These are grain regrouping, plastic deformation of the grains and bulk diffusion. The other mechanisms can only enlarge the intergranular contact surfaces and, thereby, increase the cohesion of the grain assembly. They cannot, however, affect the ratio of the void volume to the bulk volume of the material, but can only change the shape of the pores (making them more spherical).

Thümmler and Thomma (1967) distinguish the following three macroscopic stages of the sintering process, which on completion, ensure that the material obtained is solid and non-porous:

(1) an initial stage during which 'necks' grow between the neighbouring powder grains (Fig. 2.5). The molecules, however, continue to behave as 'individuals' and, thus, the grains cannot grow;

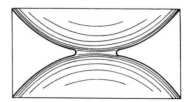

Fig. 2.5. Neck forming at the contact between two grains (after Pampuch, 1988)

(2) the stage of densification and initial growth of the grains; the molecules lose their 'individuality', the material shrinks markedly and numerous open pores are observed to occur in the material. Most grains boundaries end at the pores;

(3) this stage begins when the density of the material reaches about 90% of the theoretical density; during this stage the pores change their shapes to become more spherical and the grains continue to grow. The number of closed pores quickly increases. Densification, on the other hand, slows down. This stage is considered to come to an end when the pressure of the gases trapped within the pores exceeds the surface tension, a situation when further densification can only proceed through bulk diffusion.

The sintering techniques have long been known, but the suggestion that the major driving force involved in the densification of the ceramic material is the surface tension of the powder particles is relatively recent. Laplace has shown that, if the particles that are in contact have curved surfaces, a pressure Δp appears that tends to reduce this curvature. This pressure is given by

$$\Delta p = \gamma \left(\frac{1}{r_1} + \frac{1}{r_2} \right) \tag{2.1}$$

where γ is the surface tension of the powder particles, and r_1 and r_2 are the radii of curvature of the surfaces in contact.

The vacancy concentration gradient established by the stresses induced in the grain necks results in mass transport according to Fick's law. The movements of atoms at the contact between two grains due to diffusion are illustrated in Fig. 2.6.

As already mentioned, in different ceramic materials the sintering process proceeds in different ways and the mechanisms described above may not all occur or other mechanisms may be involved. For example, in zinc oxide, which is easily sintered, the diffusion of defects towards the grain surfaces, where they come in contact with oxygen, leads to the growth of an oxide

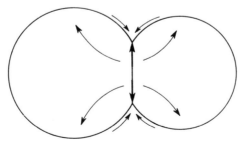

Fig. 2.6 Diffusion mass transport at the contact between two grains. The arrows indicate the direction of the atom movements: surface diffusion, diffusion along grain boundaries and volume diffusion (after Pampuch, 1988)

film there. This is an additional driving force for the process. Conversely, it is difficult to obtain high densities in materials with a great number of covalent bonds (nitrides), if the diffusion processes are inhibited.

The presence of a small amount of the liquid phase (if it does not wet the grains) during sintering does not affect the dominant role of the mechanisms described above, at least during the intermediate stages of the process. Thanks, however, to its low viscosity, the liquid phase enhances the mechanism of regrouping of the grains. As a result of the stresses acting at the grain contacts, a chemical gradient occurs there. In effect, the mass transport enhanced by the liquid phase, proceeding through dissolution and condensation, may greatly affect the kinetics of sintering.

In multicomponent systems, polymorphic transformations and chemical reactions may occur during sintering. The new phases then formed, which are more or less stable, usually greatly affect the mass transport. If this is so, the diffusion proceeds in such a direction that ensures the flow of the reactants to the reaction region. At the same time, the charge conservation principle must be fulfilled.

Let us now devote a few words to the processes of grain growth that result from the tendency for the system to reduce its free energy under the action of chemical potential gradients associated with the curved grain surfaces. A description of these mechanisms can easily be found in the literature. In multicomponent ceramic materials, the grain growth processes have usually been studied experimentally by examining the features of the materials obtained. The grain growth process may be inhibited by the presence of inclusions that block the moving grain boundaries, or it may be enhanced by the presence of the liquid phase. The grain may grow uniformly throughout the entire sample or the growth may proceed in an uncontrolled way, if single large grains grow at the expense of their small neighbours.

When considering the sintering theory, it is necessary to mention the electronic theory of sintering proposed by Samsonov (1967). This theory is based on the electronic theory of solids. By solving the Schrödinger equations, we obtain the description of a solid body in terms of the presence, the diffusion kinetics and the behaviour of point defects. The solutions of the Schrödinger equations also describe the crystal lattice energy and the structural relaxation processes. By searching deeper and deeper into the microscopic material structure, we shall certainly be able to explain the sintering mechanism more fully in the future.

Finally let us briefly describe the most recent technologies of dense materials, which will perhaps be developed in future. One is *microwave sintering* which is similar to cooking in a microwave oven (Suit Das and Curlee, 1987; Kimrev et al., 1986). Radiation of appropriate wavelength acting upon the material is absorbed throughout its whole volume, so that almost the entire radiation energy may be converted into thermal energy. Temperature gradients do not occur and, thus, the material shrinks uniformly throughout its volume. In this situation, no macroscopic stresses appear and samples of sophisticated shapes can easily be sintered in a short time. This technology provides considerable energy savings, thanks to the reduced losses and the increased sintering rate (Pampuch, 1991; Kingery, 1976).

Another example of a new technology may be *self-propagating high-temperature synthesis* (SHS), in which the material grows in a self-propagating manner (Munir, 1988). In this technique, a moving front of strongly exothermic synthesis reaction is established. After the initial activation energy is delivered, the process becomes self-propagating, since it is driven by the heat generated by the reaction. The product left behind the moving reaction front has a compact structure. The literature reports on the possibility of using this technique for fabricating almost all high-temperature non-oxide materials, such as borides, carbides, carbonitrides, sulphides, hydrides, nitrides, silicides, composites and intermetallic compounds. The materials obtained by this method have high purities and may be very dense if high pressures are employed. Considerable energy savings are possible.

In view of the emergence of still new ceramic technologies we may expect a great progress, even revolutionary developments, to take place in this field in the near future. The interested reader is referred to the literature given at the end of this chapter.

2.2 ENERGY BAND STRUCTURE OF GRAIN BOUNDARIES

This book is concerned with those ceramic materials whose properties chiefly depend on the properties of grain boundaries and thus this section provides the basic knowledge concerning the energy-band structure of grain boundaries and the mechanisms of the transport of electric current through these boundaries. It is not of course a complete description, but is sufficient for understanding the phenomena discussed later in the book.

2.2.1 Formation of a potential barrier on the surface of a semiconductor

On the free surface or at the grain boundary of a polycrystalline material, the periodic crystalline structure is discontinued. This abrupt change in the arrangement of atoms evidently disturbs the energy-band structure of the near-boundary region with respect to that of the interior of the crystal. By considering a one-dimensional model of the energy-band structure of a crystal (the Krönig–Penney model assuming the existence of square potential wells), Tamm (1932) and Shockley (1939) have proved that this discontinuity of the crystalline lattice results in surface energy states being formed. The solutions of the Schrödinger equations obtained by these authors show that the surface states are localized and are positioned within the forbidden band of the semiconductor. Many et al. (1971) have later extended the calculations to include two- and three-dimensional lattices.

In actual materials, some additional factors contribute to the formation of a surface space-charge layer. Most frequently, it is chiefly these that are responsible for the modification of the energy-band structure. Among these factors the most important are:

— an external electric field,

— the electric field induced when the material comes into contact with a material of different work function, and

— any foreign substance (even a small amount) adsorbed on the surface of the material.

2.2.1.1 Effect of an external electric field

We may examine the effect of an external electric field upon a semiconductor by, for example, examining its behaviour when it is made one plate of a condenser. We can see that the electric field alters the distribution of the carrier concentration near the semiconductor surface. Figure 2.7 shows

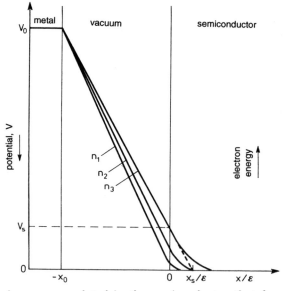

Fig. 2.7. Space charge accumulated in the semiconductor that forms a plate of a condenser charged so that its metal plate is electrically negative; $n_1 > n_2 > n_3$

how the electric field penetrates into an n-type semiconductor at three different carrier concentrations n. Since the charge on electrons is negative, their energy increases as the electric potential decreases. In order to avoid a discontinuity of the electric field at the semiconductor surface, normalized values of x/ε have been taken as a measure of the distance along the x-axis, where ε is the dielectric constant of the semiconductor. It follows from Fig. 2.7 that

$$\frac{V_s}{V_0 - V_s} = \frac{x_s/\varepsilon}{x_0} \tag{2.2}$$

If the charge density established on the plate of the condenser is $Q_s = \varepsilon_0 E$, where ε_0 is the dielectric constant of vacuum and E is the electric field intensity, and assuming that all the electrons contained within the 0–x_s region have been removed, we obtain

$$-\frac{\varepsilon_0(V_0 - V_s)}{x_0} = enx_s \tag{2.3}$$

where e is the electron charge, and n is the carrier concentration.

Equations (2.2) and (2.3) yield

$$x_s = \left(\frac{\varepsilon\varepsilon_0 |V_s|}{ne}\right)^{1/2} \tag{2.4}$$

$$\frac{V_s}{V_0 - V_s} = -\frac{\varepsilon_0(V_0 - V_s)}{\varepsilon x_0^2 ne} \tag{2.5}$$

Under the action of the electric field, a layer composed of positive ions, which are not compensated by electrons, is formed near the surface of the semiconductor.

The electrons moving towards the surface meet a potential barrier whose height V_s depends on the properties of the semiconductor, the carrier concentration n and the dielectric constant ε. Although the calculation procedure has been simplified, the results obtained are in agreement with experiment to within an order of magnitude.

2.2.1.2 Contact potential barrier

A surface potential barrier also forms when two different materials are brought into contact with one another. This effect may most easily be explained by considering an example of the contact between a semiconductor and a metal. The barrier will be the higher, the greater the difference between the work functions W_s (which represent the work that has to be done to take an electron from a material to vacuum at 0 K) of the two materials. Generally speaking, the electrons contained in the material with a smaller work function tend to flow to the material of greater work function. In the resulting state of dynamic equilibrium, a space charge builds up in the region near the surface of the former material. As a result, a potential barrier may form at the contact of such two materials, just as it does under the action of an electric field. By way of example, Fig. 2.8 shows the potential barrier formed at the contact between a metal and an n-type semiconductor whose work function is smaller than that of the metal, $W_s < W_m$. Figure 2.8a shows the energy levels, before the two materials are brought into contact. As a result of the difference between the work functions of the two materials, their Fermi levels differ by $W_s - W_m$.

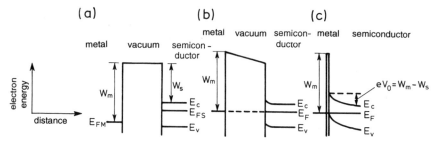

Fig. 2.8. Formation of a potential barrier at the contact between a metal and an n-type semiconductor ($W_s < W_m$): (a) the two materials are isolated from one another, (b) after reaching thermal equilibrium, and (c) after coming into close proximity

After the materials come into contact, electrons pass to the metal so as to establish a thermodynamic equilibrium and to line up the Fermi levels. In effect, a potential barrier of height V_0 (V_0 is also known as the diffusion potential) is formed such that

$$eV_0 = W_m - W_s \qquad (2.6)$$

As mentioned earlier, the charge on the electrons that have escaped to the metal is compensated by ionized donors formed at specific sites in the crystalline lattice. For this reason, the shape of the barrier and the width of the space charge region formed within the semiconductor depend on its properties. This will be discussed later.

The contact potential barrier would not form if, after the materials are brought into contact, the electrons move in the direction opposite to that assumed above (this happens when $W_s > W_m$), as shown in Fig. 2.9. If this is the case, electrons flow from the metal into the semiconductor. The Fermi level in the semiconductor is raised until an equilibrium state is reached. Because of the great carrier concentration in the metal, the space-charge region left in the metal is very narrow and the applied potential is distributed throughout the entire semiconductor. Such a contact is considered to be an ohmic contact, i.e., one that provides no barrier to the flow of current.

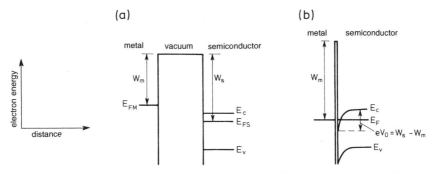

Fig. 2.9. Formation of an ohmic contact at the interface between a metal and an n-type semiconductor ($W_s > W_m$): (a) the two materials are isolated from one another, and (b) they are at close proximity

The potential barrier may of course also form at the contact between n-type and p-type regions that exist in a semiconductor (the p–n junction) and also at the contact between two different semiconductor materials (the so-called heterojunction).

The phenomena described above have formed the basis for the development of modern electronics. They are widely discussed in the literature (Kittel, 1986; Hench and West, 1990).

2.2.1.3 Surface states

Potential barriers at grain boundaries and at the surface of a semiconductor material may also form as a result of the presence of surface states. In practical ceramic materials this is often the most significant reason for their formation. Figure 2.10 shows how the acceptor-like and donor-like surface states affect the shapes of the energy bands in an n-type semiconductor. The positions of the acceptor- and donor-like states are shown before thermodynamic equilibrium is reached. The acceptor-like states below the Fermi energy are filled with electrons from the conduction band. A space charge layer compensated by the positive space charge within the semiconductor then forms on the surface. The energy bands undergo bending until all the states of energies lower than the Fermi energy are entirely filled.

The positions of the surface states within the forbidden band depend on the short-range interatomic forces, and do not depend on the so-called macropotential which describes the bending of the energy bands. This implies that during the time when electrons are filling the surface states, the positions of these states with respect to the Fermi level are changing according to the varying band bending. Figure 2.10b shows a state of equilibrium

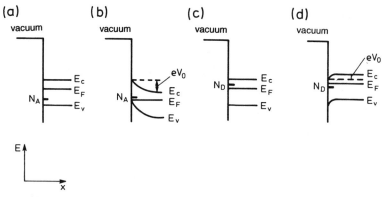

Fig. 2.10. Effect of acceptor-like (a and b) and donor-like (c and d) surface states upon the shape of the energy bands formed on the surface of an n-type semiconductor before (a and c) and after (b and d) thermal equilibrium is reached

in which the energies of partly filled surface states are higher than the Fermi energy. The same effects occur if donor-like states are present.

The acceptor-like states whose energies are greater than the Fermi energy, or the donor-like states with energies smaller than the Fermi energy, have no influence upon the shapes of the energy bands. Electrons reach the surface states at the interface between the two materials sooner than they can penetrate into the other material. The resulting distribution of the surface potential due to the surface states can predominate over the negligible contribution of the contact potential. In practice, it is often difficult to observe the effect exerted by the contact potential barrier when surface states are present, especially when their concentration is high.

If the surface of the semiconductor is ideally clean, the Tamm and Shockley surface states, mentioned earlier, occur. Under practical conditions, however, surface states appear due chiefly to external factors, such as

— adsorption of gases from the atmosphere,

— effects exerted by a liquid brought into contact with the semiconductor surface,

— adsorption of foreign atoms,

— modification of the crystalline structure due, for example, to deviations from stoichiometry.

The properties of the surface states are difficult to examine because of the problems faced in constructing appropriate model systems, which most often simulate the surface of a single crystal placed in ultrapure surroundings.

For a more comprehensive account of the nature of surface states, the reader is referred to the source literature (for example, Many et al., 1971; Kittel, 1986).

2.2.2 Shape of the potential barrier

The potential barriers formed at the grain boundaries of ceramic semiconductor materials may, to a considerable degree, determine the properties of these materials. It is not only the height of the barrier, but also its shape, dependent on the structure and the properties of the material, which is of importance here. The Gauss law describes the relationship between the distribution of the space charge, $\rho(x, y, z)$, and the electric field F produced by it, as follows

$$\nabla F = \frac{\rho}{\varepsilon\varepsilon_0} \tag{2.7}$$

where

$$\nabla = \frac{\partial}{\partial x} + \frac{\partial}{\partial y} + \frac{\partial}{\partial z}$$

The intensity of the electric field is related to the electrostatic potential $\psi(x, y, z)$ by the definition

$$F = \nabla\psi \tag{2.8}$$

$$\nabla^2\psi = \frac{\partial^2\psi}{\partial x^2} + \frac{\partial^2\psi}{\partial y^2} + \frac{\partial^2\psi}{\partial z^2} = \frac{\rho}{\varepsilon\varepsilon_0} \tag{2.9}$$

Equation (2.9) is known as the Poisson equation. Now, knowing the spatial distribution of the charge accumulated at the semiconductor surface, we can calculate the shape of the potential barrier.

Let us consider the simplest case of the one-dimensional solution of the Poisson equation subject to the boundary conditions that imply that for $x = 0$, $\psi = V_b$ (i.e., the potential at the crystal surface is equal to the height of the potential barrier), and for $x = d$ (where d is the depth of the potential barrier), $\psi = 0$. Let us further assume that we deal with an n-type semiconductor that contains no acceptors, and that the distribution of the concentration of the fully ionized donors is uniform, i.e., $N_D(x) = $ const. This situation is illustrated in Fig. 2.11. The Poisson equation then takes the form

$$\frac{d^2\psi}{dx^2} = \frac{eN_D(x)}{\varepsilon\varepsilon_0} \tag{2.10}$$

where e is the electron charge.

The solution of this equation is

$$\psi(x) = V_b - \left(\frac{eN_D(x)}{2\varepsilon\varepsilon_0}\right)x^2 \tag{2.11}$$

We can see that in this simplest case the shape of the barrier is parabolic.

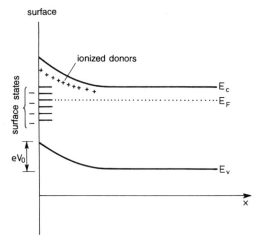

Fig. 2.11. Energy-level diagram for the surface of an n-type semiconductor with a potential barrier formed due to the transitions of electrons to surface states

Substituting $x = d$ and $\psi = 0$ gives for the height of the barrier

$$V_b = \frac{eN_D(x)d^2}{2\varepsilon\varepsilon_0} \qquad (2.12)$$

When the potential barrier occurs due solely to surface states of density N_D, the condition of electric equilibrium (charge compensation)

$$N_D d = N_S \qquad (2.13)$$

must be fulfilled. If this is so, the height of the potential barrier is

$$V_b = \frac{eN_S^2}{2\varepsilon\varepsilon_0 N_D(x)} \qquad (2.14)$$

Although these calculations have been performed for the simplest case assumed, the same procedure is often sufficiently accurate for calculating the heights of the barriers in any ceramic material. The errors introduced by it are anyway smaller than those arising from an incomplete description of the complicated structure of ceramic materials. In reality, it is improbable that the space charge distribution would be ideally uniform, that the acceptor-like levels and trap levels would be totally absent, or that these levels would be totally ionized. The solutions for more complicated cases are reported in the literature, for example by van der Ziel (1976).

2.2.3 Flow of a current through the potential barrier

As mentioned earlier, we are concerned in this book with ceramic materials of particular properties. The electrical conductivity of these materials is determined by the properties of the potential barriers formed at the grain boundaries of the matrix phase. These barriers usually reduce the electrical conductivity of the entire sample. Moreover, the electrical properties of these ceramic materials depend on external factors such as the applied voltage, temperature and ambient atmosphere. Carrier transport most often proceeds across the barrier, although in materials used for sensors it is the transport along the potential barrier which is of major importance. Below, we shall discuss some most important mechanisms of carrier transport through a potential barrier. Detailed descriptions of these phenomena were presented by, for example, Azaroff and Brophy (1963), Harrison (1980), Seager (1982), Simmons (1971), and Sze (1981).

2.2.3.1 Thermionic emission

Some theories of the conduction through potential barriers in ceramic materials have been based on the theories of thermionic emission developed by Richardson and Schottky. Let us consider the metal–semiconductor system shown in Fig. 2.12.

The condition for an electron to pass from the metal to the semiconductor is that the component of its kinetic energy, perpendicular to the potential barrier, should be greater than W'. The number of electrons able to satisfy this condition in the absence of an electric field chiefly depends on temperature. By calculating the total number of the electrons that are able to pass through the potential barrier, we arrive at the *Richardson equation* that describes the density of the thermionic emission current

$$J_s = \frac{4\pi e m_s^* k^2}{h^3} T^2 \exp\left(-\frac{E_b}{kT}\right) \qquad (2.15)$$

where e is the electron charge, m_s^* the effective mass of an electron in the semiconductor, T the absolute temperature, k the Boltzmann constant, and h the Planck constant.

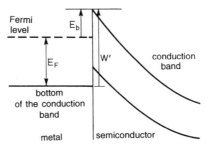

Fig. 2.12. Metal–semiconductor junction; E_F—Fermi energy, W'—the difference in energy between the highest point of the potential barrier and the bottom of the conduction band in the metal, $E_b = W' - E_F$ is the height of the potential barrier

A similar equation describes the emission of electrons from the metal into vacuum. In this case, the factor appearing before the exponential term in equation (2.15) includes the electron mass in free space (m). This factor is known as the Richardson constant, A, defined as

$$A = \frac{4\pi e m k^2}{h^3} = 1.20 \times 10^6 \text{ A/m}^2 \cdot \text{K}^2$$

Since quantum mechanics predicts that an electron may be reflected from the barrier even though its energy is sufficiently high to surmount it, the right-hand side of equation (2.15) should be multiplied by a barrier penetration coefficient, which is however usually very close to unity. In order that equation (2.15) shall include the variation of the work function with temperature, an appropriate factor representing this variation should be introduced into it.

After an appropriate rearrangement of the Richardson equation, a plot of $\ln(J_s/T^2)$ versus the reciprocal of temperature is a straight line. Such plots (known as *Richardson's plots*) are used for calculating the Richardson constant or the height of the potential barrier.

2.2.3.2 Schottky emission

Schottky found that the current density of thermionic emission depends not only on temperature but also on the intensity of an electric field applied externally (the so-called *Schottky effect*). He modified the assumption, made by Richardson, that the electrons escaping from the metal lose the energy W' in a stepwise manner. His modification consists of using the theory of the so-called image forces, which states that, after escaping from the metal, the electrons continue to be under its influence, since the Coulomb force due to the charge e^+ then induced in the metal attracts them. The work necessary for an electron to be transferred to infinity thus depends on the distance between it and the surface of the metal. Therefore, an external electric field E may facilitate the escape of electrons from the metal. Figure 2.13 shows how the potential energy $\phi(x)$ of an electron that leaves the metal varies as a function of its distance from the metal-semiconductor interface.

Schottky has showed that the density of the thermionic emission current is given by

$$J_s = J_{s0} \exp \frac{\sqrt{eE/\pi\varepsilon\varepsilon_0}}{2kT} \quad (2.16)$$

where J_{s0} is the density of this current at zero electric field intensity and may be found from

$$J_{s0} = \frac{m_s^*}{m} AT^2 \exp\left(-\frac{E_b}{kT}\right) \quad (2.17)$$

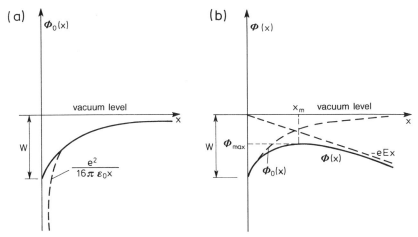

Fig. 2.13. Potential energy of the electron that leaves the metal as a function of the distance x measured from the semiconductor–metal interface: (a) without an electric field (the dashed line represents the Schottky approximation); (b) in the presence of an electric field E which produces a potential barrier with a maximum point at x_m

Here, E_b is the height of the barrier in the absence of an electric field.

The plot of $\ln J_s$ versus $E^{1/2}$ is known as a *Schottky plot*. In order to take into account the Schottky corrections to the Richardson plot, we should first draw several Schottky plots for various temperatures and find values of J_{s0} from them (by extrapolating to $E^{1/2} = 0$). Then we plot $\ln(J_{s0}/T^2)$ as a function of $1/T$. Such a procedure is often followed in calculating the heights of the potential barriers in ceramic ZnO varistors. Of course, statistics has to be involved in order to take into account the great number of potential barriers present in the ceramic material.

A characteristic feature of the carrier transport mechanisms described above is their strong variation with temperature. Generally, when carriers are transported through a grain boundary at which a potential barrier exists, the current density may be assumed to vary with temperature exponentially so that

$$J = AT^2 \exp\left(-\frac{eV_b'}{kT}\right) \qquad (2.18)$$

$$V_b' = V_b - \frac{1}{2}\left(\frac{eE}{\varepsilon\varepsilon_0}\right)^{1/2} \qquad (2.19)$$

where V_b is the height (expressed in volts) of the potential barrier in the absence of an external electric field E, V_b' is this height as modified by an electric field applied externally, and ε is the dielectric constant of the semiconductor. When $E = \mathrm{const}$, which is often the case under practical conditions, equations (2.18) and (2.19) reduce to

$$J = J_0 \exp\left(-\frac{eV_b}{kT}\right), \quad J_0 = \mathrm{const} \qquad (2.20)$$

2.2.3.3 Tunnelling

By changing the classical approach to the structure and properties of solids, quantum mechanics permitted the study and utilization of new phenomena that could not earlier be explained. One such phenomenon is tunnelling.

The solution of the Schrödinger equation, even for the simplest case describing the behaviour of an electron in a one-dimensional square potential well of finite height, shows a probability of the electron escaping out of the well. The electron can leave the well even though its energy is smaller than that required for the electrons to pass over the edges of the well. This effect is known as *tunnelling*. It has been utilized, for example, in the so-called *tunnel diodes*.

The contribution of the tunnelling effect to the current density increases with the external electric field and it becomes significant only at large field strengths (of the order of 10^9 V/m). The electric field may, for example, cause the potential barrier formed at the metal–semiconductor contact to narrow significantly, thereby greatly increasing the probability of a tunnelling transition. In principle, the transition proceeds at a constant energy. Its probability depends on the shape (the width) of the barrier and is expressed by the barrier-penetration coefficient t, which may be calculated using the methods of wave mechanics, such as the WKB (Wentzel–Kramers–Brillouin) method. Since the WKB method is quasi-classical and approximate, it may be used when the character of the particle movement is almost classical, i.e., when the de Broglie wavelength varies only slightly with distance.

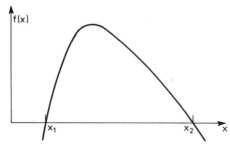

Fig. 2.14. A general case when the shape of the potential barrier is described by the function $f(x)$

If the shape of the potential barrier is described by the function $f(x)$ as shown Fig. 2.14, the coefficient representing the number of electrons that penetrate into the right-hand side of the barrier is

$$t = \varphi\varphi^* \mid_{x=x_2} = \exp\left(-2\int_{x_1}^{x_2}\sqrt{f(x)}dx\right) \quad (2.21)$$

where $\varphi\varphi^*$ is the product of the conjugate electron wave functions.

According to van der Ziel (1976), the density of the tunnelling current flowing from the metal to the semiconductor is

$$J = \frac{m_m^*}{m_s^*}\frac{e^3 E^2}{8\pi h E_b}\exp\left[-\frac{4}{3}\left(\frac{2m_s^*}{\hbar^2}\right)^{1/2}\frac{E_b^{3/2}}{eE}\right] \quad (2.22)$$

where m_m^* and m_s^* are the effective masses of the electron in the metal and semiconductor, respectively, E is the electric field intensity, and E_b is the height of the potential barrier.

We can see from equation (2.22) that the current density does not depend explicitly on temperature, but it rapidly (exponentially) increases with increasing strength of the electric field. This equation does not include several additional factors such as, for example, the forces produced by the image charge, analogous to those involved in the Schottky effect. Equation (2.22) takes various explicit forms depending upon the shape of the potential barrier and upon what additional factors are taken into account.

In ceramic materials, conduction by tunnelling does not occur in a simple form because of the complicated structure of the grain boundaries and because of its statistical character in view of the existence of many potential barriers with slightly differing heights.

One of the conduction mechanisms involving tunnelling is the so-called *Zener breakdown*, in which electrons pass from the valence band to the conduction band of the semiconductor under the action of a strong electric field. This effect may be described qualitatively in a similar way to that presented above or its description may include some other factors. Moll (1964), for example, takes into account the difference between the densities of states on the two sides of the barrier, and assumes that the barrier is parabolic in shape. A characteristic feature of the Zener breakdown is that it occurs at a relatively low voltage.

The tunnelling effect is considered to be responsible for the high nonlinearity of the current–voltage characteristic of varistors made of zinc oxide. This will be discussed later in the book.

2.2.3.4 Other effects

Among other mechanisms of conduction through a potential barrier, the one most often mentioned is *avalanche breakdown*. As a result of collisions, the electrons passing through a potential barrier in the presence of a strong electric field acquire great energy. When their energy exceeds the band-gap energy of the semiconductor, subsequent collisions generate electron-hole pairs and the number of carriers increases. The number of collisions increases in an avalanche manner. When the applied voltage reaches a critical value characteristic of a given material, the number of generated carriers increases to infinity. The current then rapidly increases and must be limited by an external circuit. The avalanche breakdown may be induced by either electrons or holes.

Contact effects and even surface effects occurring in one-element materials (Ge, Si) and in simple compounds (GaAs, GaP, InSb) whose single crystals are used in electronics have been relatively well examined. It is, however, much more difficult to describe these processes when they proceed in ceramic semiconductor materials, since here

— we deal with an assembly of many junctions formed in a polycrystalline sintered material;

— the individual junctions differ in spatial orientation, shape and dimension;

— the individual junctions show different properties because of the non-uniformity of the chemical composition, of the phase composition, and of the crystalline structure of the material;

— the purity of the ceramic materials with which we usually deal is much poorer than that of single-crystal semiconductors and thus the properties of the crystallites are more difficult to determine;

— the ceramic materials often show deviations from the stoichiometric composition;

— precipitates of foreign crystalline and vitreous phases and also adsorbed layers often occur at grain boundaries;

— under operational conditions, the contact characteristics may undergo changes due to the very presence of grain boundaries, and also due to porosity and diffusion effects.

For these reasons, the behaviour of ceramic sintered materials is, of necessity, described in a simplified manner. New theories concerning ceramic materials, even those which have relatively long been in use, continue to be developed and new technologies are still being devised. Attempts are also made to construct models that simulate individual grain boundaries and to find methods for measuring in situ the properties of the junctions formed at such boundaries. These attempts have not however yet resulted in a conclusive theory.

By appropriate technological treatment, we can produce ceramic materials with properties defined by the properties of their grain boundaries. Examples are ceramic materials used for fabricating varistors, which show nonlinear electrical conductance, ceramic materials used for fabricating thermistors whose electric conductance depends on temperature, and materials intended for fabricating sensors, with ambience-dependent electrical conductance.

2.3 SEGREGATION EFFECTS AT THE GRAIN BOUNDARIES

The preceding section was devoted to the formation of potential barriers at the grain boundaries of semiconducting ceramic materials. We discussed the simplest solutions of the Poisson equation, the equations defining the height of the potential barrier and the resulting equations describing the various carrier transport processes.

The greatest simplification which is most often made in deriving these equations lies in the assumption that the concentration of donors in the near-boundary regions is uniform. In making this assumption, the investigators are fully aware of the great effect that the distribution of donor concentration exerts upon the parameters of the potential barriers and, thus, upon the carrier transport mechanism. If we assume that the donor concentration profile in the near-boundary region is nonuniform, we can easily imagine the situation when the depth to which the space charge region of a potential barrier extends in a grain is much smaller than that suggested by the solutions mentioned above. This fact may have considerable effects, for example, on the magnitude of the tunneling current. We shall see in the chapter devoted to ZnO-based varistors how this is important in engineering applications. Surface potential, on the other hand, acting in a 'feedback loop', may in turn greatly affect the donor concentration profile. It will be remembered that this potential is established upon contact with the atmosphere (adsorption) or after a junction with another material is formed. Even when neither of these situations occurs, a small surface potential may exist (the *Tamm states*).

The surface equilibrium of a solid body, that is, of the atoms belonging to a certain region considered to be the surface, differs from the equilibrium established in the interior of the body. This is described by certain mechanisms in which the crystal is considered to consist of the interior and the surface layer, the two differing in their composition due to the tendency of the system to reach a thermodynamic equilibrium (Sachtler and Santen, 1977; Simkovich, 1982; Wagner, 1982).

The Gibbs (1948) adsorption equation derived for a two-component (1–2) system (where 1 denotes the major component) relates the excess concentration Γ_2 of the component 2 adsorbed at the surface (the excess being estimated with respect to the concentration of the component 1) to the amount of the free energy σ and the chemical potential of the dissolved substance μ_2, as follows:

$$\Gamma_2 = -\left(\frac{\partial \sigma}{\partial \mu_2}\right)_T = -\frac{1}{RT}\left(\frac{\partial \sigma}{\partial \ln a_2}\right)_T \qquad (2.23)$$

where R is the gas constant, T is temperature, and a_2 is the thermodynamic activity of the dissolved substance.

It follows from the Gibbs equation that, in a two-component solid solution, the component with a lower surface tension tends to accumulate at the surface (Haber, 1982). If the solubility of the solute is low, this component may be segregated. When the crystal comes in contact with the atmosphere, the tendency to segregation may either be increased or decreased according to the difference in the affinities of the components towards the atoms present in the atmosphere.

A similar situation occurs in those ceramic materials whose composition departs from stoichiometry, as is the case with, e.g., metal oxides. When a vacancy appears near the surface of a crystal, the crystalline lattice relaxes and the surface tension decreases.

We can, thus, see that, in accord with the laws of thermodynamics, the individual components within the near-surface region tend spontaneously to differ in their concentrations. This tendency may be enhanced by an appropriate heat treatment (Yan, 1981). This is done, for example, when fabricating PTCR thermistors and GBBL capacitors, which will be discussed later in the book. The desired distribution of atoms and point defects is obtained here by heating the material in a specified atmosphere and then cooling it down to room temperature in a controlled way. If the cooling rate is sufficiently high, the high-temperature distribution is often 'frozen in'.

Experimental examinations of the near-surface diffusion are very difficult to conduct because of the small size of the near-surface region and the difficulties associated with appropriate preparation of the surface. For example, a mechanical treatment of the surface usually introduces disturbances that result in selective gas adsorption. In the preparatory stage it is necessary to use high purity gases under high-vacuum conditions.

The techniques that allow us to study the concentration profiles in a very narrow (a few micrometres) region near the surface include the Secondary Ion Mass Spectrometry (SIMS), Auger Electron Spectroscopy (AES), and

X-ray Photoelectron Spectroscopy (XPS), among other techniques. Certain aspects of surface segregation and its influence upon the transport properties, electrical properties and reactivity are reviewed by Nowotny (1988).

A theoretical treatment of the space charge segregation in ionic solids was first proposed by Lehovec (1953) and, then, by Kliewer and Koehler (1965), Kliewer (1965), Lifshitz et al. (1967) and Poeppel and Blakely (1969). Taking into account the electrostatic interactions between charged point defects, they calculated the distributions of electrostatic potential and defects near the surface. Simple examples of pure and doped NaCl were treated. This approach was later extended to the grain boundary segregation in ceramic materials (Yan et al., 1983b,c; 1985).

Yan (1985) considered the segregation of the substance dissolved at the grain boundaries in a ceramic material. He distinguished the three major forces that drive this process:

(1) the elastic interactions with the crystal lattice due to the difference between the sizes of the ions of the solute and of the basic phase;

(2) the electrostatic interactions between the ions of the solute and the grain surface due to the presence of the space charge layer;

(3) the electrostatic dipole interactions associated with the occurrence of the point defect dipole moments.

The operational range of the forces defined in (1) is small, of the order of two lattice constants. The lattice energy increases as a square of the relative difference in the ion size. Considering potassium chloride as an example, we can see that the lattice energy increases from 0.0005 eV upon the dissolution of barium (the relative difference of the ionic radii is here $\Delta r/r = 0.015$) to 0.5 eV when magnesium is dissolved ($\Delta r/r = 0.51$). The situations becomes much more complicated when we deal with a many-element substance, such as $BaTiO_3$. Here, the electric field can deliver sufficient energy for point defects to form. The concentration profile of the point defects formed at the semiconductor surface depends upon the distribution of the electric field. Electrically charged point defects tend to agglomerate, thereby decreasing the energy of the system. The dipole moments of such defect clusters affect the field of the potential barrier.

Yan (1985), based on his earlier works (Yan et al., 1977; 1983b,c) describes how to calculate the free energy of the system. The results of these calculations confirm the tendency to segregation, mentioned earlier as justified in terms of thermodynamics. The effects of this segregation depend on the process parameters (such as the temperature, process duration and the atmosphere in which it is carried out) which should ensure the required kinetics of atom movements. This kinetics depends upon at least three factors: the rate at which the point defects form and migrate, the rate at which the solvent atoms diffuse and the rate at which the defect clusters associate and dissociate.

The parameters of the segregation process may be calculated numerically, although the calculation method is not yet fully developed. These calculations may have a great practical significance. For example, they show that boundary-layer capacitors (discussed later in the book) may be fabricated by a single sintering operation. Yan and Rhodes (1983a) describe this procedure as used for fabricating TiO_2 capacitors. Another example is the technology of low-voltage TiO_2 varistors intended for electronic applications, which has been developed based on this calculation (Yan and Rhodes, 1982a, b, c).

Recently, STEM (Scanning Transmission Electron Microscopy) examinations have confirmed the occurrence of the grain boundary segregation. By way of example, Chiang and Kingery (1989, 1990) observed the segregation of Al and O in a magnesium aluminate spinel ($MgAl_2O_4$). Observations of this kind allow us to evaluate the relative number of segregated ions. This spinel is however much more complicated than NaCl and is not purely ionic. This is why complete theoretical models are not yet available. The grain boundary segregation was also observed with Ni and Y in $SrTiO_3$, and Mn in $BaTiO_3$ (Chiang and Takagi, 1990). The latter case is of special interest when fabricating PTCR thermistors, since Mn is a common additive here that enhances the magnitude of the PTCR effect. Ikeda and Chiang et al. (1991, 1992) have recently studied the segregation phenomena in donor and acceptor doped titanium dioxide. Using the microscopic data obtained, they calculated the space charge potential and the defect formation energies. The segregation phenomena occurring in TiO_2 ceramics are in very good agreement with the space charge theory. In co-doped (Al, Nb) titanium dioxide, we observe the segregation of Al, Nb or neither of them, depending on the doping level and temperature.

2.4 ADSORPTION AND CATALYSIS

Adsorption and catalysis play very important roles during both the fabrication process of the ceramic materials dealt with in the present book, and the operation of the devices made of them. During sintering and other heat treatments involved in the fabrication of ceramic materials these two effects determine to a certain extent the properties of the grain boundaries and of the material surface. They exert a strong effect upon the operation of ceramic gas sensors. In the author's opinion, they also play an important part in the degradation of the current–voltage characteristic of ZnO-based varistors. Adsorption and the catalysis of chemical reactions are not yet fully understood, although they are utilized widely in practice. Comprehensive monographs describing these effects have been written by Volkenshtein (1987) and Ościk (1983), among other authors.

The surface of a semiconductor placed in a gaseous medium is covered with molecules or atoms of the gas, which are then adsorbed until a dynamic equilibrium is established. The kinetics of the adsorption is described by the equations of state, which interrelate the number of adsorbed molecules and time at given temperature and pressure.

The classical Langmuir (1916) theory assumes that the adsorption occurs at surface centres, separate for each molecule (one centre for one molecule) with a determined bonding energy. The gas molecules do not interact so that the energy involved in the bond with an adsorption centre does not depend on the number of molecules. The Langmuir theory describes the adsorption kinetics by

$$\frac{dN}{dt} = \alpha P(N^* - N) - \beta N \qquad (2.24)$$

where N is the number of adsorbed molecules per unit surface area, N^* is the number of the adsorption centres on this area, and P is the pressure.

The first term of this equation defines the number of molecules adsorbed on the unit surface during 1 s, and the second term represents the number of the molecules that leave this surface in the same time. In equation (2.24), α and β are given by

$$\alpha = \frac{ps}{(2\pi MkT)^{1/2}}, \qquad \beta = \nu e^{-q/kT} \qquad (2.25)$$

where M is the molecular mass, s is the active cross-sectional area of the molecule, p is the probability that, after the molecule finds itself upon a centre, it will be adsorbed, ν is the probability of the molecule being desorbed in unit time, and q is the bonding energy of the adsorbed molecule.

At the beginning of the process, when $N \ll N^*$, we have

$$\frac{dN}{dt} = \alpha P N^* \qquad (2.26)$$

Equation (2.26) is, however, not always satisfied in practice. Roginsky and Zeldowitsch (1934) describe the adsorption of CO on MnO_2 by

$$\frac{dN}{dt} = C e^{-\gamma N} \qquad (2.27)$$

where $C = $ const, and $\gamma > 0$.

For the adsorption of oxygen on ZnO and TiO_2, Stone (1955) and Thuillier (1960) give the equation

$$\frac{dN}{dt} = C e^{-\gamma N^2} \qquad (2.28)$$

By solving equation (2.26) and substituting $t = \infty$, we obtain the isotherm equation

$$N = \frac{N^*}{1 + b/P} \qquad (2.29)$$

where $b = \beta/(\alpha P + \beta)$ and $b = b_0 e^{-q/kT}$ according to equation (2.25).

When we deal with a gas mixture, each constituent gas may be described by the equation, similar to (2.29),

$$N = N^* \frac{P_i/b_i}{1 + \Sigma P_i/b_i} \qquad (2.30)$$

in which appropriate values of b_i and of the partial pressure P_i should be substituted.

Most actual systems do not, however, satisfy the assumptions of the Langmuir theory and, thus, their isothermes cannot be described by the above equations. This is due to the fact that the actual chemisorption effect is more complicated physically than that assumed by Langmuir.

Depending on the kind of forces that bind the adsorbed molecule, we can distinguish between physical adsorption, in which the bond forms without charge transfer, and chemisorption, in which this transfer does occur. In the former case, the adsorbed molecule may be acted upon by van der Waals' forces, or by forces resulting from electrostatic polarization. In both cases, activation energy may need to be delivered—we then speak of activated adsorption. The chemisorption energy depends on the nature of the chemical bond between the molecule and the crystal surface. According to the amount of this energy, we have a 'weak' or 'strong' chemisorption. For a more detailed description of the adsorption effects, the reader is referred to the literature. Here, by way of example, we shall only describe qualitatively the adsorption of oxygen on the surface of ZnO.

Three kinds of adsorption of oxygen on the surface of ZnO have been distinguished (Bonasewicz and Littbarski, 1981; Göpel and Lampe, 1980), differing from one another by the amounts of energy involved in the bonds. By heating up a ZnO sample, appropriately prepared, and measuring the oxygen partial pressure above the sample using a gauge of high sensitivity, we may determine the amount of oxygen desorbed from the crystal surface. The temperature at which the desorption reaches a maximum is proportional to the binding energy. The results of these measurements obtained by Göpel and Lampe (1980) are shown in Fig. 2.15, where we can easily distinguish the physical adsorption and chemisorption. If the sample is heated up further to a higher temperature, the oxygen bound in the crystal lattice (O^{-2}) is desorbed and then the zinc oxide begins to sublimate.

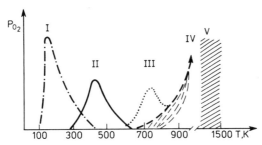

Fig. 2.15. Variation of the oxygen pressure due to thermal desorption from ZnO surface; $dT/dt = 3.3$ K·s, I—oxygen adsorbed due to van der Waals forces, II—chemisorbed oxygen, III—desorption of ZnO-bound oxygen, IV—sublimation of ZnO, V—region of high-temperature treatment (after Göpel and Lampe, 1980, by courtesy of the authors)

The physical adsorption energy ranges from 10 to 33 kJ/mol when the relative coverage of the surface decreases from 8×10^{-2} to 3×10^{-2}. The 'purely' physical adsorption is characterized by full reversibility and low reaction energy; the van der Waals forces are the only forces involved here. No other data are available in the literature (Bonasewicz and Littbarski, 1981). Chemisorption, on the other hand, has been extensively examined. Within the temperature range from 300 to 650 K, oxygen is chemisorbed as O_2^-, after removal of an electron from the conduction band. This occurs according to

$$O_{2g} \rightleftarrows O_{2s} \quad \text{physical adsorption}$$
$$O_{2s} + e^- \rightleftarrows O_{2s}^- \quad \text{chemisorption} \tag{2.31}$$

Göpel (1977) has found that the maximum coverage of the ZnO surface by chemisorbed oxygen does not exceed $\theta_{\max} = 2.5 \times 10^{-4}$, i.e., it is smaller by two orders of magnitude than that occurring in the physical adsorption.

The chemisorption of oxygen on the ZnO surface has an activation energy E_A. Lagowski et al. (1977) have proposed a model of the 'active surface states' established in the chemisorption of oxygen on ZnO. They have derived the equation defining the chemisorption rate

$$i = k_e \bar{c}_e n_{O_2} N_D e^{e_0 V_s / kT} e^{-E_A / kT} \tag{2.32}$$

where k_e is the active cross-section of an electron in the adsorbed state, $\bar{c}_e$ the thermal velocity of the conduction electrons, n_{O_2} the density of 'empty' surface states, i.e., the 'coverage' by adsorbed oxygen, N_D the density of donors inside the ZnO, $e_0 V_s$ the height of the surface potential barrier, and k the Boltzmann constant.

Physical adsorption is considered to be the initial stage of chemisorption.

In order that an electron can be taken from the ZnO, the oxygen molecule must be activated thermally and the probability that this will occur is defined by $e^{E_A/kT}$.

Lagowski et al. (1977) suggest that the chemisorption process does not depend on the crystal orientation, but other authors (Dorn and Lothe, 1977; Heiland and Kunstmann, 1969) report that it does.

When zinc oxide is heated (Fig. 2.15), the oxygen chemisorbed on its surface is desorbed at temperatures between 300 to 600 K (the desorption is maximum at 430 K). According to Göpel (1977), the desorption rate is given by

$$i_{\text{des}} = 1.4 \times 10^{13} n_{\text{ad}} e^{-110 \pm 10/RT} \tag{2.33}$$

where n_{ad} is the concentration of adsorbed oxygen, and R is the gas constant.

The adsorbed oxygen (O_2^-) molecules may react with the point defects present in ZnO and, if this is so, they dissociate. Göpel (1977, 1980) has found that, at a temperature close to room temperature, the chemisorbed oxygen reacts with oxygen vacancies or with interstitial zinc ions. At higher temperatures, the oxygen is incorporated into the lattice directly from the gaseous phase. Grunze et al. (1976) have shown that by heating ZnO in oxygen (10^5 Pa, 993 K, 24 h) we obtain oxygen-rich ZnO layers on all the surfaces of the single crystal. Since the electrons pass to the oxygen chemisorbed on the ZnO surface, a potential barrier forms. Figure 2.16 shows this situation in a n-type semiconductor. At the same time, the strong electric field (10^5–10^6 V·cm^{-1}) induced within this region may cause positive interstitial Zn_i^+ ions to pass to the surface, as a result of which a ZnO layer forms there (Bonasewicz and Littbarski, 1981; Glemza and Kokes, 1965). This is described by

$$O_{2s}^- + Zn_i^+ \rightleftarrows ZnO + O_s, \quad O_s^- + Zn_i^+ \rightleftarrows ZnO. \tag{2.34}$$

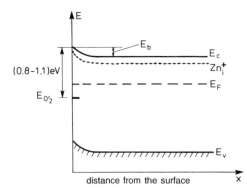

Fig. 2.16. Formation of a potential barrier on ZnO surface; E_{O_2}—the energy levels of chemisorbed oxygen, E_c—the bottom of the conduction band, E_F—Fermi level, E_v—the top of the valence band, E_b—the height of the potential barrier

The result of these reactions is the distribution of concentration of ionized donors in the surface layer as shown in Fig. 2.17.

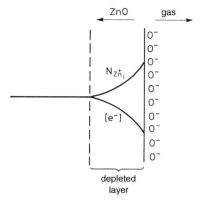

Fig. 2.17. Concentration profile of ionized Zn_i^+ donors and electrons due to oxygen adsorption on ZnO surface

The mechanisms of transport and segregation of ions within the region of the potential barrier have been studied by Yan (1985), among other investigators, and have been described in the preceding section.

At the present, it is believed that the form in which oxygen is adsorbed depends on temperature. Chon and Pajares (1969) report that at a temperature between 373 and 453 K, it is chiefly O_2^- molecules that are adsorbed, whereas above 500 K, O^- ions predominate. Kwan (1969) gives the following formula describing the sequence in which the chemisorption of oxygen varies with increasing temperature:

$$O_{2s} \rightarrow O_{2s}^- \rightarrow 2O_s^- \rightarrow 2O_s^{2-} \tag{2.35}$$

When examining the conduction and the Seebeck voltage of ZnO powders, Cimino et al. (1963) observed that with oxygen adsorbed on its surface, ZnO may even behave as a *p*-type semiconductor (*p*-type conduction has been observed very rarely in ZnO).

Catalysis effects are understood even less well than adsorption. The descriptions available in the literature are usually quantitative and refer to selected particular examples. The catalyst of a chemical reaction usually performs one of two functions (or both): it accumulates (densifies) the reagents around itself and reduces the activation energy of the reaction. The densification of the reagents occurs through adsorption or absorption of the gas particles. How the catalyst reduces the activation energy may be best illustrated by describing the reaction of hydrogen oxidation (Morrison, 1987). Let us compare the reactions of the formation of water with and without the participation of a catalyst, such as platinum, that enhances the ionization of the reagents. These reactions proceed according to

$$2Pt\cdot + H_2 \rightarrow 2Pt:H$$
$$2Pt\cdot + O_2 \rightarrow 2Pt\text{---}O \tag{2.36}$$
$$2Pt:H + Pt\text{---}O \rightarrow 3Pt\cdot + H_2O$$

$$H_2 \rightarrow 2H\cdot$$
$$O_2 \rightarrow 2O\cdot \tag{2.37}$$
$$2H\cdot + O\cdot \rightarrow H_2O$$

In the reaction defined by (2.36), the energy of bonds between the hydrogen atoms and platinum is similar to that occurring in a H_2 molecule and, thus, the energy needed for this molecule to dissociate is very small. A similar situation occurs when an oxygen molecule undergoes dissociation. In the reaction given by (2.37), on the other hand, the amount of energy required for the molecules to dissociate is much greater. This energy is returned after a water molecule forms, but the fact that it is necessary for the reaction to be activated establishes a barrier. If we, however, use platinum as a catalyst, the activation energy need not be delivered and, thanks to the instability of the Pt—H and Pt—O bonds, the hydrogen and oxygen are strongly reactive.

An ideal catalyst should not form strong stable bonds with the reagents. It should function as an agent that ensures the required course of the reaction at a reduced activation energy, whilst the reaction should not change the form of the catalyst. Catalysts may be not only metals, but also chemical compounds, including a variety of semiconducting metal oxides. The characteristic features of catalysts is their activity and selectivity. Usually, the activity of a catalyst is proportional to the increasing rate at which a given reaction proceeds in its presence. A catalyst may increase the reaction rate a hundred or even a thousand times; in the presence of a catalyst, a reaction may even proceed in an explosive way, whereas without the catalyst the same reaction may not occur at all. A good example is the oxidation of carbon oxide

$$2CO + O_2 \rightarrow 2CO_2 \tag{2.38}$$

Without a catalyst, the rate of this reaction is small and it only increases at elevated temperatures (several hundred degrees). With a catalyst, such as MnO_2, Ag_2O, Co_2O_3, it already proceeds at a high rate below room temperature (210 K). At a temperature of 270 K or more, CuO and NiO may also be used as the catalysts.

The selectivity of a catalyst is manifested in the fact that it is only active towards certain selected chemical reactions. The rate of a reaction which usually consists of several stages is determined by the rate of the slowest stage. A catalyst can change the rates of the individual reaction stages (another stage may become the slowest) or it can even make the reaction run through quite different stages. The reaction product can also be changed.

The properties of semiconductor catalysts depend on their structure; for example, they may be determined by the kind and degree of doping. The structure of a real catalyst may change during the reaction, as a result of which its activity decreases. If this is so, the catalyst is strictly associated with the course of a given reaction in which it is involved. In order to achieve its highest possible activity, this mechanism, in particular the adsorption and desorption processes that occur on the catalyst surface, should be thoroughly examined. The electron theory of the operation of catalysts, which is based on the chemistry and physics of solids, has shown that the catalytic properties of catalysts are closely related to their electrical, optical and magnetic properties. A more detailed discussion of catalysis will be given in Section 3.4, which is devoted to ceramic gas sensors. For more information the reader is referred to the literature (Kiselev and Krylov, 1987, 1989).

REFERENCES

Section 2.1

Clarke D. R. (1987), On the equilibrium thickness of intergranular glass phases in ceramic materials, *J. Am. Ceram. Soc.*, **70** (1), 15.

Coble R. L. and Burke J. E. (1963), *Sintering in Ceramics, Progress in Ceramic Science*, Vol. 3, 197, Pergamon Press, Oxford.

Dereń J., Haber J. and Pampuch R. (1977), *Solid State Chemistry* (in Polish), PWN, Warszawa.

Geguzin Y. A. (1984), *Physics of the Sintering Process* (in Russian), Nauka, Moskva.

Hsueh C. H. and Evans A. G. (1987), Liquid-phase sintering of ceramics, *J. Am. Ceram. Soc.*, **70** (10), 708.

Kimrev H. D., White T. L., Bigelow T. S. and Becker P. F. (1986), Initial results of a high-power microwave sintering experiment at ONRL, *J. Microwave Power*, **21** (2), 81.

Kingery W. D. (1976), *Introduction to Ceramics*, Wiley, New York.

Kuczynski G. C. (ed.) (1973), *Sintering and Related Phenomena, Materials Science Research*, Vol. 6, Plenum Press, New York.

Kuczynski G. C. (ed.) (1975), *Sintering and Catalysis, Materials Science Research*, Vol. 10, Plenum Press, New York.

Kuczynski G. C. (ed.) (1980), *Sintering processes, Materials Science Research*, Vol. 13, Plenum Press, New York.

Kuczynski G. C., Miller A. E. and Sargent G. A. (eds.) (1984), *Sintering and Heterogeneous Catalysis, Materials Science Research*, Vol. 16, Plenum Press, New York.

Mrowec S. (1980), *Defects and Diffusion in Solids. An Introduction*, Elsevier/North-Holland, Amsterdam.

Munir Z. A. (1988), Synthesis of high temperature materials by self-propagating combustion methods, *Am. Ceram. Soc. Bull.*, **67** (2), 342.

Pampuch R. (1976), *Ceramic Materials. An Introduction to Their Properties*, PWN, Warszawa, Elsevier, Amsterdam.

Pampuch R. (1988), *Ceramic Materials: an Outline of Nonorganic-Nonmetallic Materials* (in Polish), PWN, Warszawa.

Pampuch R. (1991), *Constitution and Properties of Ceramic Materials*, PWN, Warszawa, Elsevier, Amsterdam.

Pryadko I. F., Pryadko L. F., Timofeyeva I. I., Parhomenko V. D., Smolin M. D. and Ristic M. M. (1988), Principles and methodology of electron theory of sintering, *Sci. Sintering*, **20** (1), 7.

Samsonov G. V. (1967), Die elektronische Deutung des Sintermechanismus bei metallischen und nichtmetallischen Pulvern, *Planseeberichte für Pulvermetallurgie*, **15** (1), 3.

Samsonov G. V. and Uphahyaya G. S. (1971), Electronic mechanisms of basic technological processes in powder metallurgy of high temperature materials, *High Temperatures–High Pressures*, **3** (6), 635.

Suit Das and Curlee T. R. (1987), Microwave sintering of ceramics: Can we save energy? *Am. Ceram. Soc. Bull.*, **66** (7), 1093.

Szymański A. (1989), *Technical Mineralogy and Petrography. An Introduction to Materials Technology*, PWN, Warszawa, Elsevier, Amsterdam.

Thümmler F. and Thomma W. (1967), The sintering process, *Metallurgical Reviews*, No. 115, 69.

Uphahyaya G. S. (1988), Role of Samsonov's stable electron configuration model in sintering of real systems, *Sci. Sintering*, **20** (1), 23.

Vincenzini P. (ed.) (1983), *Ceramic Powders*, Elsevier, Amsterdam.

Wang F. F. Y. (ed.) (1976), *Ceramic Fabrication Processes; Treatise on Materials Science and Technology*, Vol. 9, Academic Press, New York.

Yan M. F. (1981), Microstructural control in the processing of electronic ceramics, *Mat. Sci. Eng.*, **48** (1), 53.

Section 2.2

Azaroff L. V. and Brophy J. J. (1963), *Electronic Processes in Materials*, McGraw-Hill, New York.

Harrison W. A. (1980), *Solid State Theory*, Dover Publ., New York.

Hench L. L. and West J. K. (1990), *Principles of Electronic Ceramics*, John Wiley & Sons, New York.

Kittel C. (1986), *Introduction to Solid State Physics*, Wiley, New York.

Many A., Goldstein Y. and Grover N. B. (1971), *Semiconductor Surfaces*, North-Holland, Amsterdam.

Moll J. M. (1964), *Physics of Semiconductors*, McGraw-Hill, New York.

Seager C. H. (1982), The electronic properties of semiconductor grain boundaries, in: Pike G. E., Seager C. H. and Leamy H. J. (eds.) (1982), *Grain Boundaries in Semiconductors*, Elsevier, Amsterdam.

Shockley W. (1939), On the surface states associated with a periodic potential, *Phys. Rev.*, **56** (4), 317.

Simmons J. G. (1971), Conduction in thin dielectric films, *J. Phys. D: Appl. Phys.*, **4** (5), 613.

Sze S. M. (1981), *Physics of Semiconductor Devices*, Wiley, New York.

Tamm I. E. (1932), Uber eine mogliche Art der Elektronenbindung an Kristalloberflachen, *Z. Physik*, **76**, 849.

van der Ziel A. (1976), *Solid State Physical Electronics*, Prentice Hall, Englewood Cliffs, N. J.

Section 2.3

Chiang Y. -M. and Kingery W. D. (1989), Grain-boundary migration in nonstoichiometric solid solutions of magnesium aluminate spinel: I. Grain growth studies, *J. Am. Ceram. Soc.*, **73** (5), 271.

Chiang Y. -M. and Kingery W. D. (1990), Grain-boundary migration in nonstoichiometric solid solutions of magnesium aluminate spinel: II. Effects of grain-boundary nonstoichiometry, *J. Am. Ceram. Soc.*, **73** (5), 1153.

Chiang Y. -M. and Takagi T. (1990), Grain-boundary chemistry of barium-titanate and strontium-titanate: I. High-temperature equilibrium space-charge, *J. Am. Ceram. Soc.*, **73** (11), 3278.

Gibbs J. W. (1948), *Collected Works*, Vol. 1., Yale University Press, New Haven, CT.

Haber J. (1982), The role of surface phenomena in determining the properties of materials, in: Nowotny J. (ed.) (1982), *Transport in Non-Stoichiometric Compounds*, PWN, Warszawa, Elsevier, Amsterdam, 349.

Ikeda J. A. S., Chiang Y. -M. and Madras C. G. (1991), Grain boundary electrostatic potential as a function of acceptor and donor doping in TiO_2 in: Mason T. O. and Routbort J. L. (eds.), *Point Defects and Related Properties of Ceramics; Ceramic Transactions*, Vol. 24, Am. Ceram. Soc., Westerville, Ohio, 341.

Ikeda J. A. S., Chiang Y.-M. and Garratt-Reed A. J. (1992), Quantification of grain boundary segregation in TiO_2 by STEM, will be published in *Proc. XIII Int. Congress on X-Ray Optics and Microanalysis*, Manchester, U.K., Aug. 31–Sept. 4, 1992.

Kliewer K. L. and Koehler J. S. (1965), Space charge in ionic crystals. I. General approach with application to NaCl, *Phys. Rev. A*, **140** (4), 1226.

Kliewer K. L. (1965), Space charge in ionic crystals. II. The electron affinity and impurity accumulation, *Phys. Rev. A*, **140** (4), 1241.

Lehovec K. (1953), Space charge and distribution of lattice defects at the surface of ionic crystals, *J. Chem. Phys.*, **21** (7), 1123.

Lifshitz I. M., Kossevich A. M. and Geguzin Y. E. (1967), Surface phenomena and diffusion mechanism of the movement of defects in ionic crystals, *J. Phys. Chem. Solids*, **28** (5), 783.

Nowotny J. (1988), Surface segregation of defects in oxide ceramic materials, *Solid State Ionics*, **28–30**, 1235.

Poeppel R. B. and Blakely J. M. (1969), Origin of equilibrium space charge potentials in ionic crystals, *Surf. Sci*, **15**, 507.

Sachtler W. M. and van Santen R. A. (1977), Surface composition and selectivity of alloy catalysts, in: Eley D. D., Pines H. and Weisz P. B. (eds.) (1977), *Advances in Catalysis*, Vol. 26, 69.

Simkovich G. (1982), Surface adsorption in multicomponent systems, in: Nowotny J. (ed.) (1982), *Transport in Non-Stoichiometric Compounds*, PWN, Warszawa, Elsevier, Amsterdam, 349.

Wagner J. B., Jr. (1982), On the role of the surface in diffusion studies, in: Nowotny J. (ed.) (1982), *Transport in Non-Stoichiometric Compounds*, PWN, Warszawa, Elsevier, Amsterdam, 340.

Yan M. F., Cannon R. M., Bowen H. K. and Coble R. L. (1977), Space charge contribution of grain boundary diffusion, *J. Am. Ceram. Soc.*, **60** (3–4), 120.

Yan M. F. (1981), Microstructural control in the processing of electronic ceramics, *Mat. Sci. Eng.*, **48** (1), 53.

Yan M. F. and Rhodes W. W. (1982a), Varistor properties of (Nb, Ba)-doped TiO_2, in: Pike G. E., Seager C. H. and Leamy H. J. (eds) (1982), *Grain Boundaries in Semiconductors*, Elsevier, Amsterdam, 357.

Yan M. F. and Rhodes W. W. (1982b), Effect of cation contaminants in conductive TiO_2 ceramics, *J. Appl. Phys.*, **53** (12), 8809.

Yan M. F. and Rhodes W. W. (1982c), Preparation and properties of TiO_2 varistors, *Appl. Phys. Lett.*, **40** (6), 536.

Yan M. F. and Rhodes W. W. (1983a), Ultra high dielectric capacitance in TiO_2 ceramics, *Adv. Ceramics*, Vol. 5, Am. Ceram. Soc., Columbus, Ohio.

Yan M. F., Cannon R. M. and Bowen H. K. (1983b), Space charge, elastic field, dipole contributions to equilibrium solute segregation at interfaces, *J. Appl. Phys.*, **54** (2), 769.

Yan M. F., Cannon R. M. and Bowen H. K. (1983c), Space charge distributions near interfaces during kinetic processes, *J. Appl. Phys.*, **54** (2), 779.

Yan M. F. (1985), Theoretical studies and device applications of solute segregation at ceramic grain boundaries, *Proc. Internat. Confer. Science of Ceramics 13*, 9–11 Sept., Orleans, France.

Section 2.4

Bonasewicz P. and Littbarski R. (1981), Adsorption phenomena, in: Kaldis E. (ed.) (1981), *Current Topics in Materials Science*, Vol. 7 (Zinc Oxide), North-Holland, Amsterdam, 371.

Chon H. and Pajares J. (1969), Hall effect studies of oxygen chemisorption on zinc oxide, *J. Catal.*, **14** (3), 257.

Cimino A., Molinari E. and Cramarossa F. (1963), Oxygen chemisorption and surface p-type behavior of zinc oxide powders, *J. Catal.*, **2** (4), 315.

Dorn R. and Lothe H. (1977), The adsorption of oxygen and carbon monoxide on cleaved polar and nonpolar ZnO surfaces studied by electron energy loss spectroscopy, *Surf. Sci.*, **68**, 385.

Glemza R. and Kokes R. J. (1965), Chemisorption of oxygen on zinc oxide, *J. Phys. Chem.*, **69** (10), 3254.

Göpel W. (1977), Oxygen interaction of stoichiometric and non-stoichiometric ZnO prismatic surfaces, *Surf. Sci.*, **62** (1), 165.

Göpel W. and Lampe U. (1980), Influence of defects on the electronic structure of zinc oxide surfaces, *Phys. Rev. B*, **22** (12), 6447.

Grunze M., Hirschwald W. and Thull E. (1976), Characterization of thin zinc-rich and oxygen-rich oxide layers, *Thin Solid Films*, **37** (3), 351.

References

Heiland G. and Kunstmann P. (1969), Polar surfaces of zinc oxide crystals, *Surf. Sci.*, **13** (1), 72.

Kiselev V. F. and Krylov O. V. (1987), *Electronic Phenomena in Adsorption and Catalysis on Semiconductors and Dielectrics*, Springer-Verlag, Berlin.

Kiselev V. F. and Krylov O. V. (1989), *Adsorption and Catalysis on Transition Metals and Their Oxides*, Springer-Verlag, Berlin.

Kwan T. (1969), Photoadsorption and photodesorption of oxygen on inorganic semiconductors and related photocatalysis, in: Hauffe K. and Wolkenstejn T. (eds.) (1969), *Proc. Symposium on Electronic Phenomena in Chemisorption and Catalysis on Semiconductors*, Moskva, July 2–4, 1968, de Gruyter, Berlin, 184.

Lagowski J., Sproles E. S., Jr. and Gatos H. C. (1977), Quantitative study of the charge transfer in chemisorption. Oxygen chemisorption on ZnO, *J. Appl. Phys.*, **48** (8), 356.

Langmuir I. (1916), The constitution and fundamental properties of solids and liquids. Part I. Solids, *J. Am. Chem. Soc.*, **38** (11), 2221.

Morrison R. S. (1987), Selectivity in semiconductor gas sensors, *Sensors and Actuators*, **12** (4), 425.

Ościk J. (1983), *Adsorption*, Horwood Int., New York.

Roginsky S. and Zeldowitsch Ja. (1934), Die katalytische Oxydation von Kohlenmonoxyd auf Mangandioxyd, *Acta Physicochim. URSS*, **1** (3–4), 554.

Stone F. (1955), *Chemistry of the Solid State*, Garner W. E. (ed.), Butterworts, London, 367.

Thuillier J. M. (1960), Contribution a l'étude de la chimisorption de l'oxygène sur l'oxyde de zinc, *Annales de Physique*, **5** (7–8), 865.

Volkenshtein F. F. (1987), *Electronic Processes on a Semiconductor Surfaces during Chemisorption* (in Russian), Nauka, Moskva.

CHAPTER 3

Metal-Oxide Varistors

A varistor is essentially a resistor having nonlinear current-voltage characteristic (the J–V characteristic) of shape shown in Fig. 3.1 (the characteristic

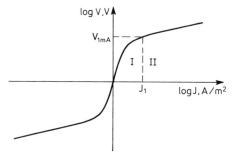

Fig. 3.1. Current-voltage characteristic of a varistor: $V_{1\,\mathrm{mA}}$ — the varistor nominal voltage measured at the current density J_1 when the current is equal to 1 mA

has been drawn on a double logarithmic scale). We can see from the figure that by increasing the voltage applied to the varistor to a certain characteristic value we can cause the current that flows through the varistor to increase rapidly. The J–V characteristic of a varistor is symmetric and may be described by the empirical equation

$$J = CV^\alpha \qquad (3.1)$$

where J is the current density (A/m^2), V is the voltage (V), $C = \mathrm{const}$, and α is a parameter, called the nonlinearity coefficient, which is characteristic of the varistor. In practice, the value of the parameter α is calculated by inserting the values J_1, J_2, V_1 and V_2 read from the J–V characteristic into the equation

$$\alpha = \frac{\log J_2 - \log J_1}{\log V_2 - \log V_1} \qquad (3.2)$$

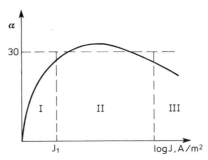

Fig. 3.2. The nonlinearity coefficient α of a ZnO-based varistor as a function of current density; at J_1, the current flowing through the varistor is equal to 1 mA

In an ideal (in terms of performance) varistor, the width of region I of the J–V characteristic should be reduced to zero (up to a specified voltage, the ideal varistor should be a perfect insulator), whereas in region II, the characteristic should run parallel to the abscissa axis (cf. Fig. 3.1). According to equation (3.1), the coefficient α will then tend to infinity. Practical varistors have, however, finite values of α. For example, the coefficient α of ZnO varistors, which will be discussed later in the book, varies with the current density in the way shown in Fig. 3.2. The varistor nominal voltage $V_{1\,\mathrm{mA}}$, indicated in the figure, is the voltage measured at a current of 1 mA. At this point, the coefficient α usually reaches its maximum. Region I, situated below the voltage $V_{1\,\mathrm{mA}}$, is called the leakage current region.

Varistors are most often used for levelling transient peaks of voltage (sudden voltage increases) which may injure the components of electric circuits.

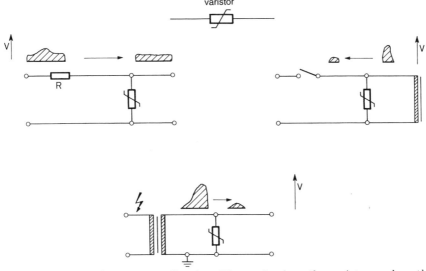

Fig. 3.3. Examples of varistor applications illustrating how the varistors reduce the effects of overvoltage in electric circuits

When such an overvoltage surge occurs, the resistance of the varistor rapidly decreases, thereby preventing the surge from reaching a protected element (Fig. 3.3). The varistor thus operates under voltage all the time, but when the voltage peaks do not occur, it acts as a passive element of very high resistance. The rated operating voltage of the varistor is usually chosen so that, on the one hand, the varistor reacts at a suitably small overvoltage, but on the other hand, its leakage current is not too large. ZnO varistors, for example, usually operate at 70–80% of their nominal voltage. When an overvoltage appears, the varistor should transmit the pulse current surge and then, as quickly as possible, return to its starting characteristic so as not to cause unnecessary current losses and not to disturb the operation of the electric circuit it protects.

Until recently, most varistors used in electronic circuits, in addition to selenium elements and Zener diodes which serve the same purpose, were fabricated from SiC ceramics. In the early 1970s, much better and more versatile ZnO-based varistors were devised and many manufacturers began to fabricate them (Matsuoka, 1981; Levinson and Philipp, 1986). For electronic applications where low operating voltages are required and the semiconductor elements employed are very sensitive to overvoltage, research work is now being carried out into the possibility of fabricating varistors of other ceramic materials, such as TiO_2.

Figure 3.4 compares the current-voltage characteristics of several varistors of various types. We can see that ZnO varistors have a large nonlinearity coefficient ($\alpha > 30$) and at the same time they can withstand high current surges. In the present book we shall chiefly deal with this type of varistor.

Varistors are now widely used in many fields, starting with electronics, where they protect diodes, transistors and similar elements, through electric power systems, where they protect the windings of transformers, motors

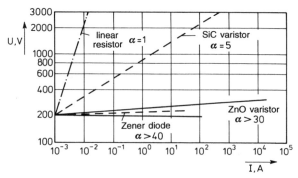

Fig. 3.4. Schematic diagram to compare the current-voltage characteristics of various varistors (redrawn from the Siemens Catalogue, 1978/79)

and all kinds of electrical contacts (Hardnen et al., 1972; General Electric, 1978; *Machine Design*, 1985; Philips, 1984; Siemens, 1978/79; 1980/81), to energy-supply equipment, where they are used as lightning arresters and transmission line protectors (Aleksandrov and Pruzhinina, 1979; ASEA, 1982; Avdeenko et al., 1976; Graciet and Salmon, 1981; Hieda et al., 1975; Kobayashi, 1978; Sakshaug et al., 1977; Snow et al., 1980b; Tominaga et al. 1979a,b, 1980). Varistors are also used in telecommunication, data processing, automobile and household equipment.

Among unconventional applications of modern ZnO varistors we can mention control devices for liquid crystal display systems (Castleberry, 1979; Levinson et al., 1982) or the protection against the effect of electromagnetic pulses during nuclear explosions (Philipp and Levinson, 1981a).

The major advantages of modern ZnO varistors can best be discussed using the example of the protection devices for electric power distribution and transmission systems. In surge arresters incorporating SiC varistors, the varistor must be connected in series with the spark gap (Fig. 3.5) and, thus,

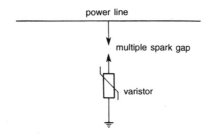

Fig. 3.5. Equivalent circuit of a surge arrester

its primary role is to restrict the delay current that flows when the plasma channel formed due to the discharge in the spark gap is slowly extinguishing. With ZnO varistors operating in transmission systems up to 500 kV, the surge arresters need not be equipped with a spark gap at all, thanks to which they become more reliable and their production and exploitation costs may be reduced.

In view of the fact that, on the one hand, overvoltage surges are more and more common in power supply systems and, on the other hand, the modern components used in various circuits, e.g., in electronic circuits, are increasingly sensitive to overvoltages, the application field of ZnO varistors may be expected to be further expanded.

In the following chapters we shall discuss the physical aspects of the fabrication of ZnO varistors, the mechanism of their operation and also similar problems associated with TiO_2 varistors. Generally, we shall confine our attention to low-voltage varistors (rated voltage up to 1000 V), but the physical phenomena occurring there may be generalized to other types of varistors. An extensive review of various properties and applications of ZnO varistors has recently been given by Gupta (1990).

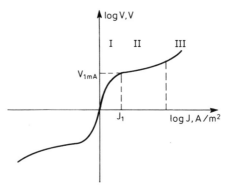

Fig. 3.6. Current-voltage characteristic of an ZnO-based varistor; $V_{1\,mA}$ — the varistor voltage; at J_1, the current flowing through the varistor is 1 mA

The current-voltage characteristic of ZnO varistors (Fig. 3.6) differs somewhat from the characteristic shown in Fig. 3.1. We can see that it includes an additional region III where the nonlinearity coefficient α again decreases.

We can mention several other parameters characteristic of ZnO varistors. All these parameters, including those discussed above, may be found in catalogues prepared by the manufacturers of these varistors. The catalogues also advise how to choose the varistor so that it fits a particular application. This choice is often more difficult to make than it is in the case of other electronic components since we must know or assess a great number of parameters.

The electrical properties of varistors are measured at an ac voltage (50 Hz) or at a dc voltage not greater than V_{1mA}. Above this voltage, current pulses of specified shape are applied. They may be rectangular shape surges 1 to 2 ms long or short 8/20 µs pulses as shown in Fig. 3.7.

Varistors are usually fabricated in the form of pellets with the electrodes deposited on their faces and the leads as shown in Fig. 3.8. The whole device is covered with a plastic material. In general, the rated voltage of a varistor is proportional to the thickness of the pellet, and the maximum admissible power increases with its diameter. The rated voltage and the diameter are indicated in the trade name of a varistor. The catalogues of varistors include tables and diagrams which indicate the number of current pulses of specified shape and magnitude that a given varistor can withstand without changing its characteristic. The admissible deviation is $+10\%$ of the varistor nominal voltage (see Section 3.5). For example, the SIOV-S14K130 Siemens varistor

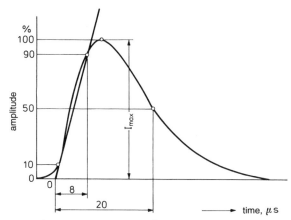

Fig. 3.7. 8/20 μs current pulse (from ANSI STD.C62.1–1979)

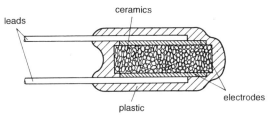

Fig. 3.8. Structure of a ZnO varistor

(see Siemens Catalogue, 1978/79 and 1980/81), whose diameter is 14 mm and rated voltage is 130 V (50 Hz), can withstand one 8/20 μs pulse of 4500 A amplitude (55 W · s absorbed energy), 10^2 500 A (4.3 W · s) pulses, 10^4 150 A (1.3 W · s) pulses, 10^6 50 A (0.3 W · s) pulses and a practically infinite number of 25 A (0.2 W · s) pulses, all of the same duration. For 2 ms surges, these values are: one 38 A pulse, 10^2 8.3 A pulses, 10^4 4.7 A pulses, 10^6 3.1 A pulses, and an infinite number of 2.5 A pulses. We can see how quickly the number of pulses that a varistor can withstand decreases with increasing amplitude and duration of the overvoltage surge applied. This behaviour results from the ability of the varistor to dissipate the generated heat and also manifests itself in its liability to degradation, which will be discussed in Section 3.5. Varistor catalogues also specify the dc varistor nominal voltage (205 V for the SIOV-S14K130 varistor just discussed) and the maximum admissible power (0.6 W for the SIOV-S14K130 varistor). The capacitance of this varistor is about 1000 pF. The catalogues also describe other varistor parameters obtained from various tests. ZnO varistors may operate at temperature between 233 and 358 K, and their customized version even up to 400 K; their temperature coefficient of the variation of the characteristic voltage is below -0.5×10^{-3} K^{-1} (the varistor nominal voltage decreases). Varistors are now fabricated by many manufacturers throughout the world, such as General Electric, Matsushita, Siemens, Philips and TDK Electronics. These companies usually manufacture a variety of varistor types; for example, in 1980, Siemens manufactured 163 different types of varistors.

The ceramic materials used for fabricating metal-oxide varistors are prepared by sintering zinc oxide in composition with certain amounts of oxides of other metals. The basic composition, devised by Matsuoka (1971) and most often used today, is

$$ZnO + 0.5 \text{ mol\%} \ Bi_2O_3 + 1.0 \text{ mol\%} \ Sb_2O_3 + \\ + 0.5 \text{ mol\%} \ CoO + 0.5 \text{ mol\%} \ MnO_2 + 0.5 \text{ mol\%} \ Cr_2O_3 \quad (3.3)$$

Varistor manufacturers also use another composition, without the addition of bismuth oxide which was initially considered to be responsible for the nonlinearity of the J–V characteristic, namely (Mukae et al., 1977; Alles and Burdick, 1991)

$$ZnO + CoO + Pr_6O_{11}$$

or $\quad(3.4)$

$$ZnO + La_2O_3 + Pr_6O_{11} + CoO$$

Bhushan et al. (1981a,b) described the ZnO + BaO system. The hundreds of patents which have been granted in this field describe many other varistor technologies which use almost all possible metal oxides. The composition and technology have not, of course, been revealed as they are protected by the patent rights, but the information available in the patent specifications suggest that the manufacturing techniques employed in the fabrication of standard performance varistors are similar to those typical of ceramics technology. A precise production regime is however usually imposed, including a high quality, predetermined composition and high purity of the raw material and also certain specified parameters of the individual technological operations. Conventional ceramic manufacturing techniques (cf. Section 2.1) include the following steps:

— weighing, mixing or grinding (in ball mills) the powder,
— calcination (if required) in air at 870–1070 K,
— regrinding,
— pressing at a pressure between 30–60 MPa (using automatic presses) to form pellets,
— sintering in air at 1370–1620 K for 1–4 hours,
— grinding the front faces of the pellets (if required),
— depositing the electrodes (Ag or Al) on the front faces of the pellets,

— heating (if required) in air or oxygen at 770–970 K,
— soldering the leads,
— depositing a protective plastic coating using, for example, the epoxy fluid bed method,
— final selection and painting appropriate symbols,
— testing the properties.

The typical composition (according to formula (3.3)) may be modified by liquid phase sintering with melted Bi and Sb oxides, whose melting temperatures are 1090 and 928 K, respectively (Samsonov, 1973). Certain reactions (discussed in Section 3.3) also occur in the solid state. A variety of works have been presented describing the dependence of varistor properties on processing conditions (Matsuoka, 1971; Asokan et al., 1987; Takemura et al., 1986, 1987; Driear et al., 1981; Hampshire and Coolican, 1987; Salmon et al., 1980).

Other techniques of varistor fabrication were also attempted, especially when extreme (low or high) values of the varistor nominal voltage were required. Snow and Cooper (1980), Kostič et al. (1987) and Asokan and Freer (1990), for example, used the hot-pressing technique, whereas Lauf and Bond (1984), the sol–gel technique. Ivers-Tiffee and Seitz (1987) and Seitz et al. (1987) studied various precursors suitable to obtain varistor-type ceramics. To reduce the varistor's breakdown voltage, we may add some large seed grains to the powder (Hennings et al., 1990). This permits us to control the grain size and, thus, to produce a homogeneous microstructure without discontinuous grain growth.

In the following chapters we shall discuss the properties of varistors fabricated using these methods (based on numerous reports available in the literature) and also the opinions of various investigators about the mechanism of electrical conduction and the degradation of the J–V characteristic of these varistors.

3.1 PROPERTIES OF ZnO AND Bi_2O

Zinc oxide is the chief component of the varistor materials, whereas bismuth oxide is considered to be one of their most important components probably responsible for the nonlinearity of their electrical conduction behaviour. Below, we shall briefly discuss the basic properties of these oxides.

3.1.1 Zinc oxide

Zinc oxide crystallizes in the hexagonal lattice of wurtzite; its lattice constants are $a = 325.0$ pm, $c = 520.6$ pm and $c/a = 1.60$. The lattice constants of zinc oxide vary depending on how great is the unavoidable deviation from stoichiometry towards an excess of the metal ($Zn_{1+x}O$). The melting temperature of ZnO is 2248 K (Samsonov, 1973); its sublimation becomes significant at as low a temperature as 1500 K. ZnO only appears in one polymorphic modification.

There are differing opinions about the type of point defects that occur in ZnO. The fact that the bonds in ZnO are 50–60% ionic (Neumann, 1981a) gives no indication as to the type of defects. The measurements of the electron density reported by Mohanty and Azaroff (1961) have shown that the dominant defects are interstitial zinc ions. Hagemark et al. (1975a, b; 1976) and Li (1975) confirm this observation. Other investigators, however, suggest that it is oxygen vacancies V_O that predominate, and this hypothesis has never been refuted. The defect equilibria, already established, have been given in the monographs by Kröger (1973/74), Sukkar and Tuller (1983), Neumann (1981a,b), and Mahan (1983).

The measurements of the content of the excess zinc in $Zn_{1+x}O$ made by Jae Shi Choi and Chul Hyun Yo (1976) have indicated that the value of x ranges from 0 to 0.07, depending on the temperature and the oxygen partial pressure; the value of x increases as the temperature increases and the oxygen pressure decreases. The activation energy of the diffusion of both zinc and oxygen in ZnO appears to be relatively small, but here also different investigators (Neumann, 1981b; Stevenson, 1973; Bonasewicz et al., 1982) give very differing values of this energy. The rate of diffusion in polycrystals is of course much greater, thanks to diffusion along the grain boundaries.

Zinc oxide is an n-type semiconductor and the carriers are generated by point defects. By analysing the equilibria of these defects, we can find that the interception point of the straight lines that represent the electron and hole concentrations versus the oxygen pressure, i.e., the point at which, at equal carrier mobilities, the n-type conduction should change into p-type

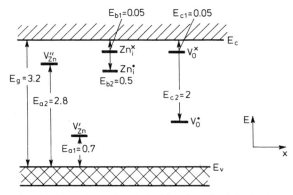

Fig. 3.9. Energy levels formed in the forbidden band of ZnO due to point defects (E in eV)

conduction, occurs at a much greater oxygen pressure than it does when the composition is stoichiometric (Dereń et al., 1977). It is also very difficult to obtain the *p*-type conduction in ZnO by varying the oxygen pressure. We can, however, greatly increase the ZnO conductivity by heating this oxide in zinc vapour.

The bandgap width in ZnO is 3.2 eV. The positions of the individual energy levels formed in the forbidden gap due to the presence of point defects are shown in Fig. 3.9. We can see that the positions of the donor levels of the interstitial zinc atoms and oxygen vacancies are equal in terms of energy.

The resistivity of ZnO ranges from 1 to 100 $\Omega \cdot$ m, which corresponds to the electron concentration of the order of 10^{21}–10^{23} m^{-3}. The Hall mobility at room temperature is about 1.8×10^{-2} m^2/V $\cdot$ s. The dielectric constant is usually taken to be $\varepsilon = 8.5$ (Hannay, 1950).

3.1.2 Bismuth oxide

According to Medernach and Snyder (1978) and Harwig and Gerards (1978), in addition to equilibrium phases α-Bi$_2$O$_3$ and δ-Bi$_2$O$_3$, two metastable phases of bismuth oxide, namely γ and β-Bi$_2$O$_3$, may also occur. The phase transformation diagram of pure Bi$_2$O$_3$ is shown in Fig. 3.10. Bismuth oxide may also occur in the form of the non-stoichiometric phases Bi$_2$O$_{2.33}$, Bi$_2$O$_{2.75}$, BiO and as a mixture of β-Bi$_2$O$_3$ with β-Bi$_2$O$_{2.5}$ (Medernach and Snyder, 1978). Levin and Roth (1964) and Safronov et al. (1970) have given two-component equilibrium diagrams for bismuth oxide and oxides of other metals.

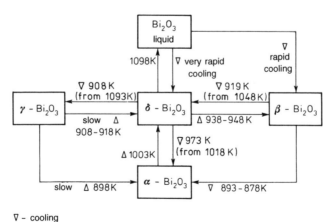

Fig. 3.10. Schematic diagram illustrating the phase transformations in pure Bi$_2$O$_3$ (reprinted from Medernach and Snyder, 1978, by permission of the American Ceramic Society)

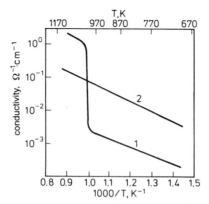

Fig. 3.11. Typical temperature variation of the electrical conductivity in pure bismuth oxide (curve 1) and when sintered with another metal oxide (curve 2)

Many authors report that the ionic conductivity (oxygen ions) of Bi_2O_3 sintered with other metal oxides is very high. A typical variation of the electrical conductivity with temperature for pure Bi_2O_3 and when sintered with another metal oxide is shown in Fig. 3.11. In pure Bi_2O_3, at a temperature of about 1000 K (during a cooling cycle), a polymorphic transformation $\delta \to \alpha\text{-}Bi_2O_3$ occurs and the conductivity decreases in a stepwise manner. When Bi_2O_3 appears in a system with another metal oxide, the high-temperature phase is stabilized and the stepwise conductivity change does not occur. Such a stabilization was observed in the following systems:

$Bi_2O_3 + WO_3$	(20–33 mol%)	(Takahashi and Iwahara, 1973),
$Bi_2O_3 + Gd_2O_3$	(35–50 mol%)	(Takahashi et al., 1975a),
$Bi_2O_3 + Y_2O_3$	(25–43 mol%)	(Takahashi et al., 1975b),
$Bi_2O_3 + M_2O_5$	(M = V, Nb, Ta)	(Takahashi et al., 1977),
$Bi_2O_3 + Sb_2O_3$	(1–10 mol%)	(Miyayama et al., 1983a),
$Bi_2O_3 + SiO_2$	($6Bi_2O_3 \cdot SiO_2$)	(Miyayama et al., 1983b),
$Bi_2O_3 + PbO$	(30–70 mol%)	(Honnart et al., 1983).

Takahashi and Iwahara (1978) suggest that this behaviour results from the stabilization of the $\delta\text{-}Bi_2O_3$ phase which has the fcc lattice structure of fluorite type. This lattice contains a great number of oxygen vacancies; this may be described by the formula $Bi_4O_6\square_2$, where $\square$ denotes the oxygen vacancy. The ionic transport numbers (the ionic-to-total conductivity ratio) of these materials, $t_i = \sigma_i/\sigma_i + \sigma_e$, exceed 0.95. Miyayama et al. (1981, 1983a) suggest that the presence of Sb_2O_3 stabilizes the tetragonal $\beta\text{-}Bi_2O_3$ phase. With the optimum 4 mol% Sb_2O_3 content, the ionic conductivity of

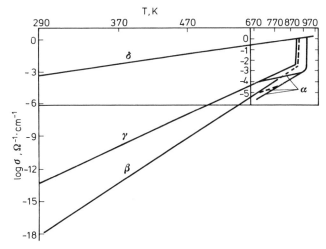

Fig. 3.12. Conductivities of various phases of bismuth oxide extrapolated to room temperature (after Einzinger, 1982a)

this material is about 2 orders of magnitude greater than the conductivity of the β phase present in pure Bi_2O_3. Harwig and Gerards (1978) examined how the conductivity of the Bi_2O_3 phases varies with temperature over the range from 673 to 1073 K. They found that, above 673 K, the conductivity of the δ-Bi_2O_3 phase is greater than that of the γ and α phases by about 3 orders of magnitude. The electron-type conductivity only predominates in the α phase. Einzinger (1982a) extrapolated these values of the conductivity of the Bi_2O_3 phases to room temperature (Fig. 3.12) and found that the conductivity of the δ-Bi_2O_3 phase at this temperature is greater by 9 orders of magnitude than the conductivity of the γ phase. Similar results may be derived from data given by other investigators.

3.2 ELECTRICAL PROPERTIES OF ZnO VARISTORS

3.2.1 Current-voltage characteristics

The shape of the current-voltage characteristic of a ZnO varistor has been described earlier in this chapter (Fig. 3.6). The varistor nominal voltage V_{1mA} increases linearly as the varistor thickness increases, and it decreases linearly as the average size of the ZnO grains decreases. An increase in the temperature of the sintering process, enhancing the growth of ZnO grains, results in a decrease of the varistor nominal voltage (Wong, 1976; Brückner, 1980; Hozer, 1984, 1985; Adveenko, 1978; Milosevič et al., 1983). These facts were the first to suggest that the nonlinear behaviour of this ceramic material

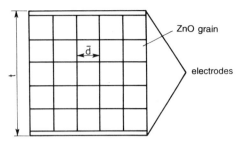

Fig. 3.13. Model of the structure of a varistor-type material with cubic ZnO grains

is associated with the effects that occur at the boundaries of the ZnO grains. The first model of the microstructure of this material was then constructed. The model was very simplified but permitted calculation of the electrical properties of the individual grain boundaries, once the properties of the whole sample had been measured (Fig. 3.13). The properties of the grain boundaries calculated in this way are often in surprisingly good agreement with measured values. For example, if t is the thickness of the sample and $\overline{d}$ is the average diameter of the ZnO grains, then

$$s = \frac{t}{\overline{d}} \qquad (3.5)$$

is the number of junctions along the sample thickness, and

$$V = \frac{V_{1\,\mathrm{mA}}\overline{d}}{t} \qquad (3.6)$$

is the voltage across a single grain boundary when the voltage applied to the sample is equal to the varistor nominal voltage. The accuracy of these calculations is not of course very high, since in reality the average grain diameter is a statistical parameter (Emtage, 1979; Hozer and Szymański, 1983).

The current-voltage characteristic of a varistor varies with temperature in the way shown in Fig. 3.14 (Philipp and Levinson, 1979). In region I (regions I, II and III are indicated in Fig. 3.6), the effect of temperature is relatively strong and the current quickly increases with increasing temperature. In regions II and III, the effect of temperature weakens considerably. The voltage $V_{1\,\mathrm{mA}}$ only slightly decreases with temperature. Hence, we can infer that the carrier transport mechanisms, predominating in the leakage region and in regions II and III, differ from one another.

The current-voltage characteristic of a varistor depends evidently on the factors involved in its fabrication. This characteristic is nonlinear only when the sintering process is carried out in air or in oxygen (Matsuoka, 1971). According to Salmon (1980), a reduction in the oxygen partial pressure leads

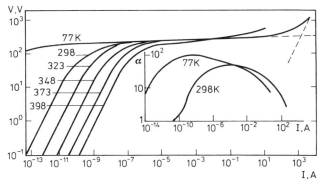

Fig. 3.14. Variation of the current-voltage characteristic of a ZnO varistor with temperature

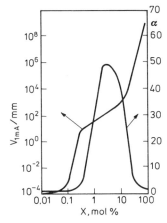

Fig. 3.15. Variations of $V_{1\,mA}$ and α as functions of the total content of oxide additives; X — the total content of Bi_2O_3, Sb_2O_3, CoO, MnO_2 and Cr_2O_3 (1:2:1:1:1). Reprinted from Matsuoka (1971) by permission of *Japanese Journal of Applied Physics*

to an increase in the leakage current and to a rapid decrease of the nonlinearity coefficient α. Matsuoka (1971) examined how the voltage $V_{1\,mA}$ and the nonlinearity coefficient α depend on the total content of oxide additives (Fig. 3.15). He found that, although the varistor nominal voltage increases with increasing content of additives, the nonlinearity coefficient reaches a maximum when this content is a few per cent. Since, as has already been mentioned, the voltage $V_{1\,mA}$ also depends on the sintering temperature, we may expect that the varistor properties can be controlled by adjusting the process parameters. Using the same chemical composition of the material and fabricating pellets of the same thickness, we can produce varistors of different types and different nominal voltages. This is important since the ability of a varistor to absorb the energy of a current surge depends on its volume.

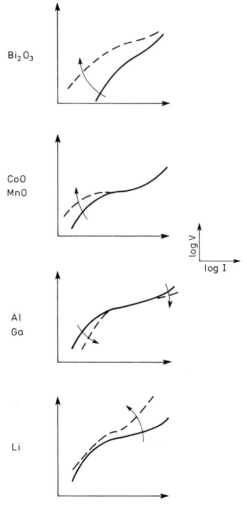

Fig. 3.16. Effect of oxide additives upon the shape of the J–V characteristic of ZnO varistors (after Eda, 1982)

Figure 3.16 shows how an addition of various metal oxides affects the J–V characteristic of a varistor. It is believed that an addition of Bi_2O_3 results in the nonlinear behaviour of the varistor. Other investigators suggest that the same role may be played by oxides of other metals of large ion radius, such as rare earth metals (Mukae et al., 1977; Williams et al., 1980; Mukae, 1987) or barium (Bhushan et al., 1981a,b). Small ion radius metal oxides, such as Mn and Co, diffuse into the ZnO grains and also increase the nonlinearity of the J–V characteristic. They also affect the conductivity of the ZnO grains in such a way that region III of the J–V characteristic (see Fig. 3.6) shifts towards greater current densities. The shape of this characteristic within region III depends directly on the ZnO resistivity, but it most strongly depends on the content of oxides of trivalent metals of small ion radius, such as Al and Ga (Carlson and Gupta, 1982; Okuma et al., 1983; Takemura et al., 1983; Miyoshi et al., 1981). The amount of the latter additives, should not, however, exceed 50–150 ppm, since they greatly increase the leakage current.

It should be noted that the leakage current also increases when the varistor is subjected to a hydrostatic pressure (Gupta et al., 1977) or when the frequency of the current flowing through it increases (Levinson and Philipp, 1976a).

3.2.2 Dielectric properties

The dielectric constant of ZnO-based varistor-type material is much greater than the dielectric constants of its constituents. It ranges from 1000 to 2000 and increases linearly as the thickness of the varistor increases. The capacitance of a varistor increases with increasing temperature, but decreases by about 20–30% with an increase of the dc voltage applied to the varistor, provided that this voltage is below the varistor nominal voltage. When an ac voltage greater than $V_{1\,\mathrm{mA}}$ is applied at a frequency below 10 kHz, the capacitance of the varistor rapidly increases with increasing voltage (Einzinger, 1982a; Morris, 1973, 1976; Mukae et al., 1979; Hozer and Szymański, 1983; Matsuura and Masuyama, 1975; Matsuura and Yamaoki, 1977). At higher frequencies, the capacitance decreases as the applied voltage increases (Einzinger, 1982a; Lou, 1980).

Having measured the variation of the varistor capacitance as a function of the applied dc voltage (the C–V characteristic), we can calculate the

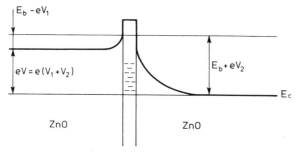

Fig. 3.17. The Schottky barriers formed at the boundary of a ZnO grain; $V = V_1 + V_2$ — the applied voltage, E_b — the height of the potential barrier at $V = 0$ (after Mukae et al., 1979)

parameters of the potential barriers at the ZnO grain boundaries. To calculate the C–V characteristic, Mukae et al. (1979) used the model of the grain boundary energy structure shown in Fig. 3.17 (such models will be discussed in more detail in Section 3.4). In this figure, a junction is subjected to the voltage $V = V_1 + V_2$. If the donor concentration N_D is assumed to be uniform, the solution of the Poisson equation for this case gives for the junction capacitance C_j:

$$\frac{1}{C_j} = \left(\frac{2}{e\varepsilon\varepsilon_0 N_D}\right)^{1/2} \left[(E_b - V_1)^{1/2} + (E_b + V_2)^{1/2}\right] \qquad (3.7)$$

If we assume that the change in the barrier height at its left-hand side is negligibly small as compared with the change at its right-hand side ($V_1 \ll V_2$), that is

$$V = V_1 + V_2 \approx V_2 \qquad (3.8)$$

then equation (3.7) takes the form

$$\left(\frac{1}{C_j} - \frac{1}{2C_{j0}}\right)^2 = \frac{2}{e\varepsilon\varepsilon_0 N_D}(E_b + V) \qquad (3.9)$$

where C_{j0} is the capacitance of the junction at $V = 0$.

A plot of $(1/C_j - 1/2C_{j0})^2$ versus V is a straight line, the inclination of which permits us to calculate the donor concentration N_D. The point of intersection of the extension of this straight line with the V-axis indicates the height of the potential barrier E_b. When passing from the single junction to the macroscopic scale of the whole specimen, we should take into account the adopted varistor model, such as described by equations (3.5) and (3.6). Hence we obtain for the varistor capacitance

$$\frac{1}{C} = \frac{s}{AC_j} \qquad (3.10)$$

where A is the surface area of the specimen, and s is the number of junctions across its thickness. It should be noted that this approximation may introduce a significant error when calculating the parameters of a single barrier, since the assumption that the grains of a ceramic material are cubic is very rough.

The total capacitance of the varistor is given by

$$\left(\frac{1}{C} - \frac{1}{2C_0}\right)^2 = \frac{2s^2 E_b}{A^2 e\varepsilon\varepsilon_0 N_D} + \frac{2sV}{A^2 e\varepsilon\varepsilon_0 N_D} \qquad (3.11)$$

Figure 3.18 shows the experimental C–V characteristics and the plots of equation (3.11) as obtained by Hozer and Szymański (1983). The parameters

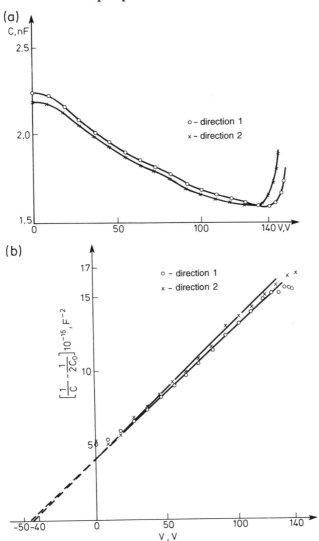

Fig. 3.18: (a) Variation of the capacitance of ZnO varistor as a function of the dc voltage applied to it; (b) plot of equation (3.11) for the same varistor as in (a) (after Hozer and Szymański, 1983)

of the specimen were measured in both directions. By inserting the typical values: $A = 1.24 \times 10^{-4}$ m^2, $s = t/\overline{d} = 10^{-3}$ m/(16.5×10^{-6}) m and using Fig. 3.18 and equation (3.11), we obtain $N_D = 6.73 \times 10^{23}$ m^{-3} and $E_b = 0.75$ V.

Hozer and Szymański (1983) also examined the variation of the J–V characteristic with temperature and constructed the Schottky plot (cf. Subsection 2.2.3) shown in Fig. 3.19. The measurements were performed at

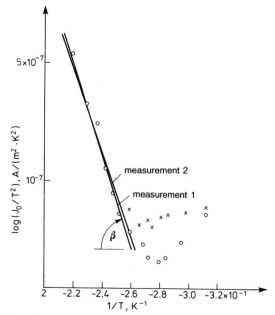

Fig. 3.19. The Schottky plot for a metal-oxide varistor; the measurements taken over the temperature range from 300 to 450 K (after Hozer and Szymański, 1983)

decreasing (measurement points 2) and increasing (measurements points 1) temperature. From the inclination of these straight lines, defined by the angle β, we can calculate the height of the potential barrier as

$$\tan\beta = \frac{eE_b}{k} \rightarrow E_b = \frac{k\tan\beta}{e} \tag{3.12}$$

In Hozer and Szymański's measurements, E_b was 0.44–0.50 eV. The very fact that the measured points lie on a straight line indicates that the Schottky effect occurs. The deviations observed at low temperatures are due most probably to measurement errors.

The variations of the varistor capacitance (C) and of the dielectric loss coefficient ($\tan\delta$) with frequency are shown in Fig. 3.20. We can see from this figure that, at room temperature, $\tan\delta$ reaches a maximum at about 10^5 Hz; at a temperature of 200 K, its maximum shifts to about 10^2 Hz (Levinson and Philipp, 1976a, b). Gupta et al. (1977) have also found that the capacitance of a varistor decreases with increasing hydrostatic pressure.

Studies on the dielectric properties of varistors have finally proved that potential barriers do exist at the boundaries of the ZnO grains, and that it is thanks to these barriers that ceramic materials show nonlinear electrical conductivity.

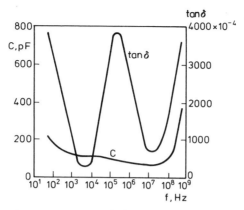

Fig. 3.20. Variation of the capacitance (C) and the dielectric loss coefficient ($\tan \delta$) of a varistor with frequency

Many investigators measured the properties of what are known as microvaristors, i.e., the single grain boundaries occurring in ceramic materials (van Kemenade and Ejinthoven, 1975, 1978, 1979; Bernasconi et al., 1976; Einzinger, 1975, 1979; Lou, 1979). This was usually done by coating the polished surface of a sintered material with a net of electrodes placed on neighbouring ZnO grains that were separated by an actively conducting boundary or by a boundary that contained an intergranular phase. These measurements have delivered much valuable information about the operation mechanism of ZnO varistors, but all quantitative calculations based on them suffer considerable errors, primarily because of the difficulty in evaluating the boundary area that is involved in the measurement. When measuring, for example, the electrical conductivity, the error results from an inaccurate evaluation of the current density.

3.2.3 Parameters of the potential barriers

The nominal voltage across a single grain boundary was determined using the linear variation of $V_{1\,\mathrm{mA}}$ and C with the thickness of the varistor and inserting a calculated average value of the diameter of the ZnO grain. The characteristic voltage thus obtained by various investigators was: 1 V (Matsuoka, 1971), 3.4 V (Wong, 1976), 2 V (Morris, 1976), 3.3 V (Mahan et al., 1978, 1979), 1.44–1.55 V (Mukae et al., 1979; Tsuda and Mukae, 1987). Hozer and Szymański (1983) have shown that, depending on the way in which the average ZnO grain boundary is calculated, the differences in the calculated nominal voltage of a single barrier may be significant. The values that they obtained ranged between 1.73 and 2.95 V for the same material. The calculated $V_{1\mathrm{mA}}$ also depends on the distribution of ZnO grains in the ceramic material: the presence of a number of large grains may greatly alter the results of calculations (Brückner et al., 1980; Emtage, 1979; Bowen and Avella, 1983).

The magnitudes of the varistor nominal voltage obtained by examining the microvaristor properties and using statistical techniques were: 4 V (van Kemenade and Ejinthoven, 1975), 3.5 V (van Kemenade and Ejinthoven, 1978, 1979), 2.3–2.9 V (Bernasconi et al., 1976), 3.2–3.6 V (Einzinger, 1975, 1979, 1982a), and 2.3–3.0 V (Lou, 1979) for ZnO–Bi_2O_3 junctions produced by ion sputtering. These values slightly exceed those obtained from the measurements of the properties of the varistor examined as a whole.

Olsson and Dunlop (1987, 1989b, c, 1993) describe four types of ZnO interfaces: A—a boundary containing a thin (2 nm) intergranular Bi-rich film, B—a ZnO–ZnO homojunction with segregated Bi atoms, C—the interface between ZnO and the intergranular mixture of Bi_2O_3 and spinel grains, and D—the interfaces between ZnO and pyrochlore, and between ZnO and spinel grains (the microstructure of ZnO varistors is described in details in Section 3.3). The J–V characteristics which they measured at the individual types of interface differed from one another. Boundaries of the type A and B had symmetrical J–V characteristics with a breakdown voltage of 3.2–3.6 V. The characteristics of C-type boundaries were asymmetrical with a breakdown voltage of 3.2 V for the electrons travelling from ZnO into Bi_2O_3 and of 0.4–0.9 V for the electrons travelling in the opposite direction. D-type boundaries were not examined. Differences between various types of grain boundaries were also reported by Tao et al. (1987) and Dorlanne and Tao (1987). Olsson and Dunlop (1989c) and also Einzinger (1975, 1978b) showed that varistors quenched from above 1170 K do not exhibit nonlinear properties. There is a certain optimum cooling rate at which the varistor has the best properties. According to Olsson et al. (1989a), the interfacial barriers begin to form during cooling from temperatures between 1070 and 1270 K.

Van Kemenade and Ejinthoven (1975, 1978, 1979) and Einzinger (1975, 1978a, 1982a) have shown that $V_{1\,mA}$ may even vary along a single grain boundary, because of its heterogeneity or because of the presence of an intergranular phase. The height of the potential barrier formed at a ZnO grain boundary ranges from 0.5 to 1.0 eV. It should be noted that the barrier heights calculated using the parameters that have been measured for the varistor as a whole may also be greatly in error for the same reasons as those discussed above in reference to the values of $V_{1\,mA}$.

Mahan (1984) has reported that the height of the potential barrier at a grain boundary may fluctuate along this boundary by a value of 0.1 eV (which is considerable as far as the electrical conductivity is considered). Such fluctuations may result from a nonuniform distribution of the electric charge that contributes to the formation of the potential barrier. Because of these fluctuations, the averaged barrier height determining the capacitance of the varistor may differ from the barrier height that actually determines the electrical conductivity.

The concentration of donors in ZnO grains of the varistor-type ceramic material has been reported to be (in m^{-3}): 3×10^{22}–3×10^{23} (Knecht and Klein, 1978), 1.3×10^{23} (Emtage, 1977), 0.76–1.6×10^{24} (Mukae et al., 1979), 3.5×10^{23} (Philipp and Levinson, 1976a), 1.1×10^{24} (Hozer and Szymański, 1983). For a model single-crystal junction, Schwing (1980, 1981, 1984) obtained 6.3×10^{22}. For the sake of comparison the donor concentration in pure ZnO was reported to be 2×10^{23} (Hannay, 1950). As, however, mentioned in Section 3.1, these values depend strongly upon how the ZnO composition departs from stoichiometry and, in varistors, also upon the composition of the material. Unfortunately, no data are available on the distribution of the donor concentration within the space charge region near the potential barrier, which, as discussed in Section 2.3, probably occurs there and may affect significantly the results obtained. An increase of the carrier concentration increases the leakage current that flows through the varistor.

3.2.4 Other properties

At low temperatures below 273 K, slowly vanishing polarization currents have been reported to occur in varistors (Philipp and Levinson, 1976b; Modine et al., 1990a; Modine and Wheeler, 1990b). Philipp and Levinson (1981b) have found that the minimum time necessary for the varistor to begin conducting in a nonlinear manner (after the applied voltage exceeds a specified value) can be as short as 5×10^{-10} s. But the results of this measurement vary widely depending on the parameters of the current pulse applied to the varistor, on the varistor capacitance and on the parameters of the measuring circuit. The pulse response of the varistors was extensively studied by Modine et al. (1987, 1989, 1990b). Eda (1979a, b) also observed current oscillations in varistors.

These results, obtained by examining the electrical properties of varistor-type ceramics, have been used for constructing models of the energy structure of the electrically-active ZnO grain boundaries and also models of the carrier transport over these boundaries. These models will be discussed in Section 3.4.

3.3 MICROSTRUCTURE OF ZnO VARISTORS

In order to describe the operating mechanism of ZnO-based varistors and the role played by the various oxide additives, it is essential to know the microstructure of the material, the distribution of the additives and the kinds of phases formed. Since the shape of the current-voltage characteristic of a varistor is determined by the properties of the near-boundary regions in the ZnO, we shall pay special attention to the character of these regions and discuss the disputed occurrence of the intergranular phase.

3.3.1 Two-component systems

The nonlinear behaviour of ZnO-based ceramic materials was first discovered in two-component systems, such as $ZnO-Bi_2O_3$ (Matsuoka et al., 1969; Kosman and Petcold, 1961). Characteristically, this nonlinearity was due to metal oxides whose ionic radii were greater than that of Zn^{+2}. These oxides show a marked tendency to segregate at the boundaries of ZnO grains. Bismuth oxide later became one of the most important components of highly-nonlinear materials, and many investigators consider its presence to be crucial for this nonlinearity to occur. The melting temperature of Bi_2O_3 is 1090 K. Hence, during the process of sintering at a temperature between 1370 and 1620 K, it forms a liquid phase, which, on cooling, crystallizes at the boundaries of the ZnO grains.

In accord with the $ZnO-Bi_2O_3$ phase diagram, these two oxides do not dissolve in one another (Levin and Roth, 1964; Safronov et al., 1970). This diagram has however been constructed for specimens in which the minimum Bi_2O_3 content was 3 mol%. In such a composition the microstructure of the phase boundary regions cannot be determined unequivocally.

By subjecting a sintered $ZnO-Bi_2O_3$ (0.5 mol%) specimen to selective etching in a $HClO_4$ solution, Wong (1974) isolated the bismuth oxide phase, which he considered to be an intergranular phase that surrounded almost the entire surfaces of the ZnO grains so as to separate them from one another. He identified this phase as crystalline β-Bi_2O_3 with no ZnO atoms. However, zinc could have been etched out, and this suggestion was confirmed by Morris (1973) who examined the intergranular Bi_2O_3 phase and found 16–33% of ZnO in it. Also Stanisic (1984) suggests that a small amount of Bi_2O_3 can dissolve in ZnO.

Wong (1980) carried out systematic studies on the sintering of the $ZnO-Bi_2O_3$ system. He found that an addition of bismuth oxide greatly increases the densification rate of ZnO due to the involvement of the liquid phase in the sintering mechanism discussed earlier. The maximum density was obtained at a linear shrinkage of 15% and a temperature of about 1170 K. Above this temperature, the density began to decrease. The addition of a small (up to 2 mol%) amount of bismuth oxide increased the size of the as-sintered ZnO grains. It also increased the porosity of the sintered material, from submicrometre values to 2–12 µm. Above a temperature of 1470 K, the mass of the sample decreased appreciably, chiefly because some Bi_2O_3 had evaporated during the sintering process.

Morris (1976) measured the values of the dihedral angle between ZnO and the intergranular Bi_2O_3 phase and stated that these two phases are not likely to wet one another. Hence it follows that bismuth oxide in a liquid phase cannot spread out freely over the surfaces of ZnO grains and, thus, the intergranular phase must be discontinuous. For a two-component system, this has been confirmed by Kingery et al. (1979) and Chiang et al. (1982), who observed the boundaries of ZnO grains by means of a high-resolution, scanning transmission electron microscope (STEM) with a magnification of 500 000 times. They found that discontinuous Bi-rich precipitations, with a non-zero dihedral angle, occur at the grain boundaries, but an approximately 20 nm wide portion of the grain boundary between them is enriched with Bi. The maximum ratio of the number of Bi to Zn atoms in this region was about 0.04 (Fig. 3.21). The same authors suggest that a small number of Bi

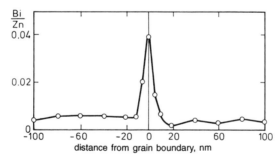

Fig. 3.21. Bi concentration at the boundary of a ZnO grain within the region where no other phases are present; the composition of the ceramic material was: ZnO+0.5 mol% Bi_2O_3 (redrawn from Kingery et al., 1979, by permission of the American Ceramic Society)

ions may diffuse into the ZnO lattice and take substitutional positions. This may be one of the factors responsible for the occurrence of potential barriers (which is in fact in accord with the charge conservation principle in view of the difference between the valencies of Zn^{+2} and Bi^{+3}). We cannot, however, exclude the possibility that even a thin Bi film adsorbed on the ZnO surface, if appropriately prepared, may produce surface states there. Unfortunately, investigators are not as yet able to examine directly the monoatomic films adsorbed on the surface of a ceramic material.

Interesting results have been reported by Kutty and Raghu (1989), who fabricated highly nonlinear varistors in the ZnO–Cu system. Ceramics containing a fraction of mol% Cu, uniformly distributed, showed electrical properties similar to those of common varistor compositions. The varistor effect was attributed to the acceptor-like behaviour of Cu.

3.3.2 Multicomponent systems containing Bi_2O_3

Just as in the case of two-component systems, most investigators studying multicomponent systems believed initially that an intergranular phase surrounded the ZnO grains in a continuous way. This opinion was first expressed by Matsuoka (1971), who is now considered to be the discoverer of modern ceramic varistor-type materials. He attributed the nonlinear conductivity of ceramics to the presence of an intergranular phase (of an average thickness of the order of 1 µm), which contained all the components of the ZnO–CoO–Bi_2O_3–Sb_2O_3–Cr_2O_3–MnO system. Co, Mn and Cr can, in addition, diffuse into ZnO grains, forming a homogeneous solution within the ceramic material. As the temperature of the sintering process increases, the size of the ZnO grains increases; in the multicomponent system specified above, subjected to sintering at 1620 K, the average grain size is about 10 µm. The variation of the ZnO grain size with temperature and duration of the sintering process have been described by Hozer and Kołacz (1984, 1985) and by Wong (1976, 1980) (Fig. 3.22). Trontelj et al. (1983) report that the ZnO grain size may also be greatly increased by adding a certain amount of TiO_2.

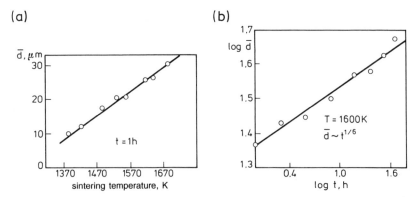

Fig. 3.22. Variation of the average diameter of the ZnO grain in a multicomponent varistor-type material as a function of (a) sintering temperature, and (b) sintering time t (after Wong, 1976)

Wong (1975) examined the properties of the phases that occur in sintered varistor-type materials. The way in which this material absorbs 500–2000 nm radiation indicates that a substitutional solution of Co^{+2} ions (about 1 at %) occurs in the ZnO lattice. The ZnO grains, which form the basic phase of the ceramics, are oriented in the space in a random way and often twinned. The characteristic green colour of the material is due to the diffusion of Co ions. According to Philipp and Levison (1975), the presence of Co ions results in deep trap energy levels being formed 2 eV below the edge of ZnO conduction band. At the boundaries of the ZnO grains, crystals of regular shapes occur. They have the fcc structure of the II–V $Zn_7Sb_2O_{12}$ or $Zn(Zn_{4/3}Sb_{2/3})O_4$ spinel phase, with lattice constant $a_0 = 860.1 \pm 0.1$ pm, and they are enriched with Mn and Co. Between the ZnO grains, the fcc phase of pyrochlore ($Bi_2Zn_{4/3}Sb_{2/3}O_6$) occurs with lattice constant $a_0 = 1048 \pm 2.0$ pm; this phase is also enriched with Mn and Co. To examine these phases, the ZnO was etched away selectively in an aqueous solution of $HClO_4$ (1:1 by vol.). The results of this examination suggest that these phases also contain a certain number of Zn ions. As the sintering temperature increases, the likelihood of the reaction

$$Bi_2(Zn_{4/3}Sb_{2/3})O_6 \xrightarrow{ZnO} Zn(Zn_{4/3}Sb_{2/3})O_4 + Bi_2O_3 \qquad (3.13)$$

increases. This reaction increases the pyrochlore/spinel ratio, as confirmed by experiment. Wong did not, however, find free Bi_2O_3 phases in the microstructure. He stated that the ZnO grains were surrounded by a pyrochlore layer, about 500 nm thick, which played the role of the intergranular phase responsible for the non-ohmic conductivity of the ceramic material. As the sintering temperature was increased from 1370 to 1670 K, the density of the ceramic material decreased from 5.46 to 5.27 g/cm^3 and its porosity increased from 2 to 6%. Wong attributes this decrease in density to the fact that at the sintering temperature Bi_2O_3 evaporates.

Wong et al. (1975) then found that the intergranular phase that occurs at the boundary between two ZnO grains is amorphous and that its composition is similar to that of pyrochlore identified earlier. Graciet et al. (1978, 1980) also reported crystals of spinel structure (of composition similar to that of $Zn_{2.1}(Co_{0.2}Sb_{0.7})O_4$) and other crystals of pyrochlore structure ($Bi_{1.4}Mn_{0.4}Zn_{0.7}Sb_{1.5}O_7$) that occur at the ZnO grain boundaries. Above 1270 K, the pyrochlore phase is decomposed to produce liquid Bi_2O_3. At higher temperatures, a new pyrochlore phase, $Bi_{0.6}Mn_{0.4}Sb_{1.8}Zn_{1.2}O_7$, which Wong calls P_2, appears; it is only stable at a high temperature (about 1570 K). On cooling, this phase is again decomposed to produce spinel crystals and free bismuth oxide, which mostly evaporates, but its remnants suffice to form a continuous intergranular layer around the ZnO grains.

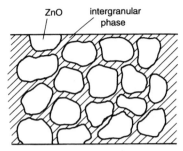

Fig. 3.23. Microstructure of a varistor-type ZnO-based material with a continuous intergranular phase. Crystals of spinel structure have been neglected as electrically-inactive

So far, in describing the microstructure of ZnO-based varistor-type materials, we assumed the occurrence of an intergranular phase that surrounds the ZnO grains in a continuous manner (Fig. 3.23). Santhanam et al. (1979), however, provide evidence that most of the zinc oxide grains touch one another without any intermediate intergranular phase. Similar hypotheses were already formulated earlier by Morris (1976) and Levine (1975). Santhanam et al. (1979) examined the varistor-type material using indirect (selective etching) and direct (by means of a transmission electron microscope) methods. They found that the intergranular phase usually occurs at the contacts between three ZnO grains, but they could not find any intergranular layer thicker than 2.5 nm at the boundaries between two such grains, even at a microscope magnification of 100 000 times. Neither did they find the amorphous phase described by Wong et al. (1975). Santhanam et al. (1979) have confirmed the frequent occurrence of twinned ZnO grains.

By examining the ceramics using an electron microprobe and a scanning electron microscope, Hozer et al. (1983, 1987) confirmed the observation that the intergranular phase tends to occur at the boundaries of three ZnO grains and that there are grain boundaries at which this phase does not occur. Figure 3.24 shows a microphotograph of the microstructure of a varistor-type material obtained by these investigators. At a high magnification, the intergranular phase is clearly visible even though the surface has been polished (Fig. 3.24b), but it does not occur at all the boundaries. Hozer et al. (1983, 1987) have also shown that Bi ions cannot dissolve in the ZnO grain structure.

Clarke (1978, 1979) examined the microstructures of the GE-MOV™ varistors manufactured by General Electric, using advanced transmission electron microscopy and high-resolution X-ray techniques. The results that they obtained for three neighbouring ZnO grains confirmed those described above, except that some evidence appeared that amorphous phases might occur in this type of varistor. Clarke used statistics to examine the occurrence of the intergranular phase at about 100 boundaries between two adjacent ZnO grains and found this phase in a few cases only. All the boundaries containing the intergranular phase were ideally straight lines along several μm and their planes were perpendicular to the basic plane of the hexagonal ZnO grain lattice. The thickness of the intergranular layer was constant along the whole boundary and was determined by the magnitude of the surface tension. Clarke also found that, whether or not the intergranular phase occurred, was dependent upon the morphology of the nearest contact between three ZnO grains and, thus, upon the crystallographic orientation of the two-grain contact. The dihedral angles between the intergranular phase and ZnO, measured at many (several tens) points, ranged from 12 to 85°. At grain boundaries containing a continuous intergranular layer, the angle was equal to zero, which could suggest that this orientation was favoured. These facts explain the observations made by Morris (1976) that the shape of the current-voltage characteristic of a varistor does not change even though the content of the intergranular phase varies over a wide range. This phase is placed at the contacts between three ZnO grains and, in terms of statistics, the number of the two-grain boundaries occupied by this phase is constant and so is the number of boundaries where it does not occur. The boundaries free of the intergranular phase were only slightly enriched with Bi, which is in accord with the results obtained by Kingery et al. (1979) and Morris (1976). The observed concentrations of Bi indicate that it does not form a discrete layer, but is rather adsorbed at the boundaries.

The wetting of ZnO grains by a Bi-rich intergranular phase was studied thoroughly by Gambino et al. (1989). They found that the wetting increases with increasing temperature. The measured dihedral angle was 0° at 1410 K and above 55° at 880 K. When the varistor was quenched from a high temperature, the Bi-rich phases remained at almost all ZnO grain boundaries, but when it was cooled slowly, these phases receded to the three-grain junctions leaving the ZnO–ZnO homojunctions behind.

Iga et al. (1976a) reported the occurrence of δ- or β-Bi_2O_3 phases in varistor-type ceramics. After heating the material at a temperature of 970 K, these phases underwent polymorphic transformations into the γ-Bi_2O_3 form. These results have been confirmed by Hozer (1988) who examined the material by X-ray diffraction.

The microstructure of ZnO-based ceramics doped with oxides such as Bi_2O_3, Sb_2O_3, Co_2O_3, Cr_2O_3, and MnO_2, and its formation during the sintering and cooling cycles were extensively studied by Inada (1978a,b; 1979, 1980, 1983). He examined the individual phases both in situ in the varistor-type material and in systems that were synthesized of individual

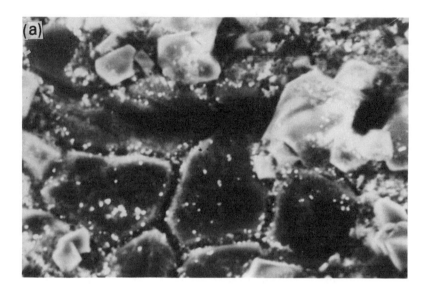

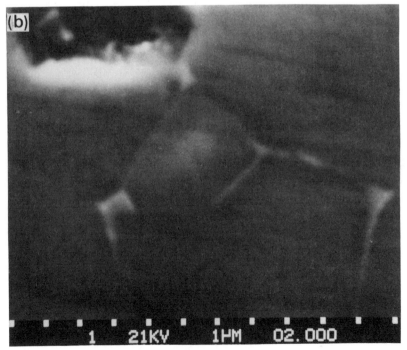

Fig. 3.24a, b

Sec. 3.3] Microstructure of ZnO varistors

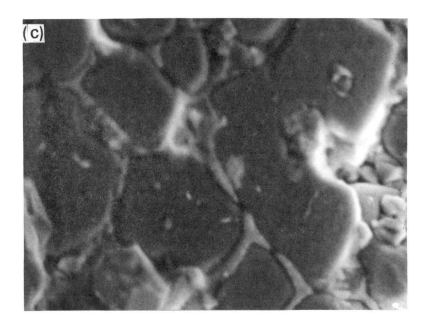

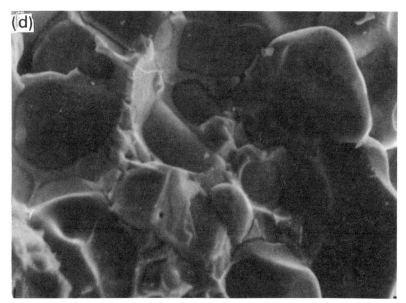

Fig. 3.24. Microstructure of a varistor-type material: (a) the free surface after sintering (magnified 3200 times); (b) the surface after polishing (magnified 10 000 times); (c) the surface after polishing and thermal etching at 1220 K (magnified 1800 times); (d) the surface of a fracture (magnified 1800 times)

constituents. To determine the phase composition at various values of X and various temperatures, he used specimens with the composition (Fig. 3.25)

$$(100 - X)\text{ZnO} + \tfrac{1}{6}X(\text{Bi}_2\text{O}_3 + 2\text{Sb}_2\text{O}_3 + \text{Co}_2\text{O}_3 + \text{MnO}_2 + \text{Cr}_2\text{O}_3) \quad (3.14)$$

Inada states that the nonlinearity of the current–voltage characteristic of a varistor increases when the phase transformations $B \to C \to D$ or $D + B'$ occur. Like Wong (1975) in formula (3.13), Inada reports that, when heated, pyrochlore reacts with ZnO according to

$$2\text{Zn}_2\text{Bi}_3\text{Sb}_3\text{O}_{14}(\text{Cr}) + 17\text{ZnO} \xrightarrow{\text{heating}} 3\text{Zn}_7\text{Sb}_2\text{O}_{12}(\text{Cr}) + \\ + 3\text{Bi}_2\text{O}_3(\text{Cr}) \text{ (liquid)} \quad (3.15)$$

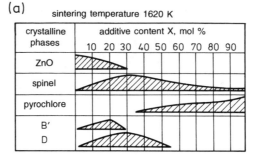

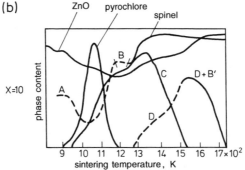

Fig. 3.25: (a) Phase compositions of various samples of a varistor-type ceramic versus X (according to equation (3.14)) at a sintering temperature of 1620 K; (b) at various sintering temperatures and $X < 30$: ZnO–ZnO with Co and Mn additives, spinel — $\text{Zn}_7\text{Sb}_2\text{O}_{12}$ added with Co, Mn and Cr, pyrochlore — the $\text{Zn}_2\text{Bi}_3\text{Sb}_3\text{O}_{14}$ phase of the pyrochlore type chiefly containing Bi, Sb and Zn with Co, Mn and Cr additives; A — Bi-rich phase; B — phase of the δ-Bi_2O_3 type, with composition close to $12\text{Bi}_2\text{O}_3\text{Cr}_2\text{O}_3$; C — tetragonal phase close to $14\text{Bi}_2\text{O}_3\cdot\text{Cr}_2\text{O}_3$; D — tetragonal β-Bi_2O_3 containing a considerable amount of dissolved Zn and a small amount of Sb; B' — phase of the δ-Bi_2O_3 type formed in the Bi_2O_3–ZnO–Sb_2O_3 system, which forms a solid solution with the phase B (reproduced from Inada, 1978a, by permission of *Japanese Journal of Applied Physics*)

In the systems without Cr_2O_3 the spinel phase does not form, but its polymorphic form known as X_{ZSO} occurs instead. At higher temperatures, this polymorphic form is transformed into the spinel phase. If the cooling operation proceeds slowly, the reaction reverse to (3.15) may occur and pyrochlore is formed. The spinel phase is stabilized (no reverse reaction) by oxide additives in the following sequence:

$$0 < Co < Mn \leq Cr + Mn \leq Cr < Co + Mn + Cr \qquad (3.16)$$

The decomposition of the spinel is also controlled by the growth of its grains.

During sintering, the Bi_2O_3 liquid phase is formed due to reaction (3.15), provided that the mole ratio $Bi_2O_3/Sb_2O_3 < 1$. If the mole ratio is greater than unity, bismuth oxide melts at a temperature of about 1070 K.

The process of formation of Bi_2O_3-rich phases strongly depends on the composition of the material and on the sintering conditions. The β-Bi_2O_3 phase, formed after fast cooling, may contain great amounts of dissolved ZnO (up to 25 mol%) and a small amount (of the order of 3 mol%) of Sb_2O_3. Inada reports that, at a slow cooling rate, the composition of this phase is close to $78Bi_2O_3 + 19ZnO + 3Sb_2O_3$.

The δ-Bi_2O_3 phase usually forms when cooling proceeds at higher rates. Inada suggests that the δ- and β-Bi_2O_3 phases form a continuous intergranular phase that surrounds ZnO grains. If this is the case, the compositions and structures of these phases determine the nonlinearity of the ceramics, thanks to which this property can be controlled by adjusting the sintering conditions. This opinion clearly departs from, for example, that of Clarke. Nevertheless, Inada's works are considered to be the fundamental contribution to the study of the formation and phase composition of varistor-type ceramics.

3.3.3 Multicomponent systems with addition of rare earth metal oxides

Mukae et al. (1977, 1981, 1987) studied the microstructure of another group of varistor-type ZnO-based ceramics doped with oxides of rare earth metals, such as

$$ZnO + 0.1 \text{ at\% Pr} + 5.0 \text{ at\% Co}$$
$$ZnO + 0.5 \text{ at\% La} + 0.5 \text{ at\% Pr} + 1.0 \text{ at\% Co}$$

These materials show nonlinearity similar to that of the ceramics that contains an addition of Bi_2O_3. Using a scanning electron microscope, Mukae et al. examined the intergranular phase that surrounded the ZnO grains. They isolated this phase by etching the material selectively in the HF–HNO$_3$–HCl–H$_2$O mixture. The intergranular phase was found to be the crystalline hexagonal La$_2$O$_3$–Pr$_2$O$_3$ solid solution, probably free of Co and Zn. They also observed an increased concentration of Co ions uniformly distributed within the ZnO grains.

Alles and Burdick (1991) have recently also observed an eutectic liquid phase present at grain boundaries in varistors doped with praseodymium oxide.

Williams et al. (1980) studied the microstructure of the grain boundaries in the varistor-type ceramics manufactured by TDK Electronics. The composition of the material was ZnO + 10 mol% CoO + 0.8 mol% BaO + +0.2 mol% NdO + 0.25 mol% Sm$_2$O$_3$. Unlike Mukae et al., they did not find any intergranular phase around the ZnO grains. The oxide additives (except CoO) were grouped in the form of small precipitates at the grain boundaries. They did not occur inside the ZnO grains. Examination using a transmission electron microscope confirmed that no intergranular layer thicker than 1 nm was present. These results were not surprising since, in view of the high melting temperatures of the oxides of rare earth metals, they should not appear in the liquid phase during the sintering process.

3.3.4 Microstructure of grain boundaries

The microstructure of varistor-type ceramics was also studied by constructing models of microvaristors (i.e., the isolated grain boundaries) contained in a polycrystalline material and of single junctions produced using thin film techniques.

R. Einzinger studied extensively the microvaristors isolated at the individual ZnO grain boundaries in a polycrystalline ceramic material. Many investigators observed grain boundaries which are not surrounded by any intergranular layer, but which, however, behave in a nonlinear way. This property was also shown by microvaristors (provided that the specimens had been prepared properly).

Baumgartner and Einzinger (1981) proposed an interesting hypothesis, which explains the differences between the structure of ZnO–ZnO boundaries in pure, sintered zinc oxide that show a linear behaviour and the structure of such boundaries in a varistor-type material doped with Co, Bi and Sb. They performed the following experiment. A powdered mixture of Bi$_2$O$_3$ and the Zn$_7$Sb$_2$O$_{12}$ spinel was placed between polished (ZnO+CoO) plates, and heated at 1270 K. According to a reaction reverse to (3.15), pyrochlore and zinc oxide are then produced on cooling. The zinc oxide is deposited on the ZnO+CoO plate, forming an epitaxial layer. Since the temperature is too low

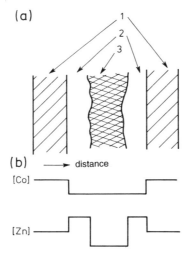

Fig. 3.26: (a) Epitaxial ZnO layer obtained by Baumgartner and Einzinger (1981): 1 — ZnO+CoO substrate, 2 — epitaxial ZnO layer, 3 — crystallized pyrochlore; (b) distribution of Co and Zn concentration within the sample

to enable the CoO contained in the substrate to diffuse quickly into the layer, it is possible to evaluate the thickness of the deposited ZnO layer. The results of this experiment are shown schematically in Fig. 3.26. Baumgartner and Einzinger suggest that such 20 nm thick epitaxial layers (of stoichiometric composition) whose structure differs from that of the substrate (which are the ZnO grains), may be formed in varistor-type ceramics, and that it is these layers which are responsible for the formation of the potential barriers and for the nonohmic conductivity of varistors. A question that remains to be explained is why pyrochlore has not been found at the electrically-active ZnO boundaries.

Measurements of the properties of the microvaristors formed in a polycrystalline material were also made by van Kemenade and Ejinthoven (1978, 1979). Using electron microscopy and electron beam induced current (EBIC) techniques, they have shown that a ZnO–ZnO boundary, even though it is not surrounded by any intergranular phase, may be inhomogeneous along its length. The effect of the electron beam upon the magnitude of the active current varies from place to place along the grain boundary.

Schwing and Hoffmann (1981) and Hoffmann and Schwing (1981) examined a model system, in which a layer composed of a mixture of the oxides Bi_2O_3, MnO_2, Co_3O_4, Sb_2O_3 and Cr_2O_3 placed between two ZnO single crystals was sintered at 1223 K. The current–voltage characteristic of the junction thus produced was nonlinear, and the nonlinearity coefficient α exceeded 15. They also attempted to produce low-voltage varistors (with

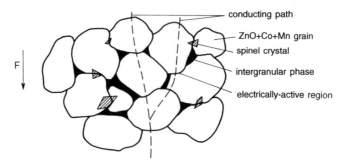

Fig. 3.27. Microstructure of a varistor-type ZnO-based material; F — the direction of the external electric field

nominal voltage from 3.2 to 3.5 V) by sintering a similar layer placed between two sintered ZnO pellets subjected to axial or hydrostatic pressure. The nonlinearity coefficient of the varistors thus obtained amounted to 19.

These various opinions about the microstructure of varistor-type ceramics permit us to offer the following description (Fig. 3.27). The basic phase of the material is composed of semiconductor ZnO grains that contain dissolved Co and Mn ions. Most of these grains touch one another without any other phase located between them, forming double ZnO–ZnO boundaries (homojunctions). The structure of these boundaries is modified due to the presence of atoms of great ionic radii, such as Bi atoms. Electric current is conducted just through these boundaries and for this reason they are referred to as the electrically-active boundaries. At the contacts between three ZnO grains, and sometimes also between two grains (perhaps those of a particular orientation), an intergranular phase occurs, most probably inactive in the conduction process. This phase chiefly contains bismuth oxide in various polymorphic forms, or its compounds such as pyrochlore. All constituents of the ceramic material dissolve in this intergranular phase. Under sintering conditions, crystals with the spinel structure also form at the grain boundaries, but they do not participate in the conduction of electric current.

The existence of conducting paths has recently been confirmed experimentally. Hohenberger et al. (1991) proposed several methods for observation of these paths. In the galvanic method, the surface of a varistor acted as the cathode of a galvanic cell in contact with $CuSO_4$ and $AgNO_3$. Metals deposited on the conducting areas of the surface and the conducting paths were visualized. In another experiment, a metal point tip was moved across the surface (polished, without the electrode) of the varistor. The current from a constant current source was passed through the tip and the voltage variations across the varistor were monitored. Several parallel linear scans can show the voltage distribution. The distribution of conducting paths was also visualized in a scanning electron microscope. Greuter and Blatter (1990) describe electroluminescence observations of the conducting paths at the grain boundaries in a varistor operated in the breakdown region.

3.4 MECHANISMS OF ELECTRICAL CONDUCTION IN ZnO VARISTORS

Based on the results obtained from studies on the electrical properties and microstructure of metal-oxide varistors, presented above, we may formulate the problems that a model of their electrical conduction should explain. These are:
— the high values of the nonlinearity coefficient α,
— the effect that various ambient factors exert upon the current-voltage characteristic of the varistors,
— the role played by the intergranular phase,
— the contributions of the individual elements and phases contained in the ceramic material to the nonlinear behaviour of these varistors,
— the importance of the individual parameters of the fabrication process.

It has been stated that the potential barriers established at ZnO grain boundaries play the decisive role in the nonlinear conductivity of varistors. The shape and the temperature variation of their J–V characteristic suggest that at least two carrier transport mechanisms predominate here, one below and the other above the nominal voltage of the varistor.

The earliest models of a varistor reported in the literature assumed that the ZnO grains are surrounded by a continuous intergranular phase. Matsuoka (1971) suggested that the current flowing through a grain boundary (the space charge induced current) is constrained by space charge and affected by a great number of trap levels present in the intergranular phase. This current rapidly increases when the voltage becomes sufficiently high for the carriers injected into the intergranular phase to increase in number so as to exceed the number of carriers generated by the traps.

Later experiments showed that, at voltages below the nominal value $V_{1\,\text{mA}}$, the relationships characteristic of thermionic emission over the Schottky barrier, described earlier, are fulfilled. It has been found that the relationship $C^{-2}(V)$ is rectilinear and that the Schottky equation (see Subsection 2.2.3) is satisfied.

Levinson and Philipp (1975a,b; 1978a,b), Philipp and Levinson (1977) and Klein and Strassler (1975) proposed a model of the conduction of a metal-oxide varistor which utilized, in a simple manner, the effect of Schottky's emission at low voltages, and the tunnelling effect at voltages above the nominal value. They assumed that potential barriers exist inside the

ZnO grains and that these grains were surrounded by a thin intergranular non-insulating layer. The density of the Schottky emission current (through a single barrier) may be calculated from

$$J = J_0 \exp\left(-\frac{E_b - \beta F^{1/2}}{kT}\right) \qquad (3.17)$$

where $\beta = (e^3/4\pi\varepsilon\varepsilon_0)^{1/2}$, and F is the intensity of the electric field.

At voltages higher than the varistor nominal voltage, the effect of electron tunnelling is described by the empirical relationship

$$J = \frac{3.39 \times 10^{-2} F^2}{E_b} \exp\left(-\frac{7.22 \times 10^7 E_b^{3/2}}{F}\right) \qquad (3.18)$$

The tunnelling effect does not depend directly on temperature, but the number and energies of the electrons reaching a potential barrier vary with temperature. Each of these factors introduces a small negative temperature coefficient of the varistor nominal voltage, which was confirmed by experiment.

Eda (1978) suggests that two back-to-back Schottky barriers exist at ZnO grain boundaries, and that they are separated from one another by an intergranular layer whose width does not exceed 50 nm (Fig. 3.28). Electrons are injected into the intergranular layer (into the surface states of the ZnO conduction band) due to thermionic emission (at low voltages) or field emission (at high voltages). The electrons then tunnel through localized trap levels. Having reached the surface states on the other side of the layer, the electrons tunnel again (by field emission) into the ZnO conduction band (at high voltages) or overcome the barrier by thermionic emission (below the nominal voltage value $V_{1\mathrm{mA}}$).

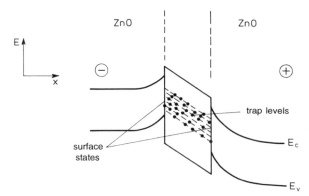

Fig. 3.28. Energy structure of the ZnO grain boundary at $V > 0$; E_c — bottom of the conduction band, E_v — top of the valence band (reproduced from Eda, 1978, by permission of the American Institute of Physics)

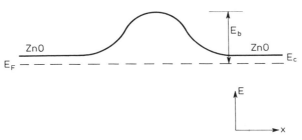

Fig. 3.29. Energy structure of the ZnO grain boundary at $V = 0$ (after Vandanamme and Brugman, 1980)

Vandanamme and Brugman (1980) assume that the only effect of the presence of Bi_2O_3 and other additives in a varistor-type material is that, at the ZnO grain boundaries, the energy bands are curved (Fig. 3.29), but no separate intergranular layer occurs there. Vandanamme and Brugman also assume that at low voltages the paths along which the carriers are transported through the varistor are longer than those calculated by dividing the specimen thickness by the average grain size (hence it follows that the paths are curved—the percolation paths). Electrical conduction is then achieved by thermionic emission over potential barriers. The current increases due to the action of a barrier lowering factor, which is in this case stronger than that when Schottky's emission is considered. The current (of the whole varistor) is given by

$$J = pAjA^*T^2 \left(\frac{m^*}{m}\right) \exp\left(-\frac{E_b}{kT}\right) \exp\left(\frac{\gamma V^{1/4}}{kT}\right) \quad (3.19)$$

where J is the current, pAj the total surface area (smaller than that of the electrodes) through which the current is conducted, γ an empirical constant, $A^* = 1.2 \times 10^6$ A/(m^2·K^2) the Richardson constant, m^* the electron effective mass, m the electron mass at rest, and E_b the initial height of the barrier (at $V = 0$). The last term in equation (3.19) represents the barrier-lowering factor.

The validity of equation (3.19) has been confirmed by experiment. The relationship $\ln J$ vs. $V^{1/4}$ has been taken to be rectilinear.

At high voltages, the conducting paths are shortened (become more rectilinear) and Zener emission takes place according to the equation

$$J = KF^2 \exp\left(-\frac{4(2m^*)^{1/2}E_g^{3/2}}{3e\hbar F}\right) \quad (3.20)$$

where $K = $ const, and E_g is the width of the ZnO bandgap.

Bernasconi et al. (1976) have, however, proved that the direct transition from thermionic emission to tunnelling cannot explain the high values of the nonlinearity coefficient observed in practice. Later works usually mentioned additional effects, due to which the current increased after the nominal voltage was exceeded. These effects include the lowering of the bottom of the ZnO conduction band on one side of the grain boundary so that it is positioned below the top of the valence band on the other side of the boundary. This effect is expected to occur at voltages above 3.2 V across a single potential barrier.

Bernasconi (1977) and Knecht and Klein (1978) assume that, at the ZnO grain boundaries, a thin (< 10.0 nm) intergranular layer exists and that it induces surface states on the interface. This in turn leads to the formation of space charge layers in the near-boundary ZnO regions and, thus, to the formation of potential barriers there. Figure 3.30 shows a model of the energy structure of a boundary at the moment when the bottom of the conduction band in one grain and the top of the valence band in the neighbouring grain are equalled. At low temperatures (below 100 K), the conduction proceeds according to the equation

$$J = AT^2 \frac{E_b}{kT} \exp\left(-\frac{\Delta}{kT}\right) \int_0^\infty dE \exp\left(-\frac{EE_b}{kT} - \frac{E_b}{C_0 kT_0} g(E, V_g)\right) \quad (3.21)$$

where: $kT_0 = \hbar/2(\gamma N_D/m^*)$, $V_g = V/V_b$, and $V_b = E_b/e$.

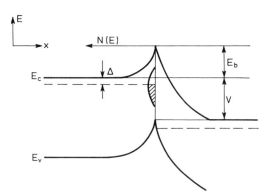

Fig. 3.30. Structure of the ZnO grain boundary; $V = V_{1\,\mathrm{mA}}$, $N(E)$ denotes the concentration of surface states (reproduced from Knecht and Klein, 1978, by permission of the Ceramic Forum International, Berichte der Deutschen Keramischen Gesellschaft)

The carrier transport proceeds by hopping over the localized states established within the potential barrier and by thermionic emission. The multiple tunnelling through localized states is described by the tunnelling factor C_0 ($C_0 \geq 1$). The function $g(E, V_g)$ is determined by the shape of the barrier. The situation occurring at voltages above $V_{1\,\text{mA}}$ is shown in Fig. 3.30. The electrons may tunnel from the ZnO valence band directly to the conduction band in an adjacent grain, leaving behind localized states which may then be filled again through recombination. Bernasconi (1977) assumes that the tunnelling occurs from deep surface states, an effect which in terms of mathematics is equivalent to that just described. The recombination, however, probably proceeds very slowly. If this is so, an inversion layer forms within the junction region and, as the voltage is further increased, the height of the potential barrier decreases according to equation (3.22) (at voltages higher than that at which the bottom of the conduction band on one side of the boundary is aligned with the top of the ZnO valence band on its other side):

$$V_b(V) = V_b(0) \quad V \quad (3.22)$$

This effect is manifested by an increase of the hopping current due to the reduction of the barrier height. It is assumed that with increasing temperature the contribution of thermally-induced conduction increases.

Hozer and Szymański (1984) suggest that an analogy exists between the mechanism of the sudden current increase, mentioned above, and the switching mechanism observed in some glasses and crystals.

Hower and Gupta (1979) suggest that the nonlinearity of the J–V characteristic of varistors is a function of the distribution, in terms of both energy and space, of the surface states over the interface between two phases. As the voltage exceeds the nominal voltage of the varistor, the charge collected in surface states increases and the height of the potential barrier decreases. At a certain voltage, a sudden increase of the current occurs. Emtage (1977) suggests that this sudden increase in current takes place at the moment when all the surface states present in the intergranular phase become filled.

Mahan et al. (1978, 1979, 1982) assume that the holes created near the boundary of a ZnO grain also contribute to this sudden current increase. Between two Schottky's barriers, a thin intergranular layer with trap states of high density exists. At low voltages, conduction is due to thermal activation, but above a certain critical voltage V_c, which Mahan defines as

$$eV_c = E_g - E_b = 3.2 - 0.8 = 2.4 \text{ eV} \quad (3.23)$$

where E_g is the ZnO bandgap and $E_b = 0.8$ eV, and if the barrier height is constant, holes begin to be created, as illustrated in Fig. 3.31. They form an additional space charge which alters the shape of the potential barrier so that it becomes thinner. The tunnelling current in the ZnO conduction band then greatly increases and becomes the predominant conduction com-

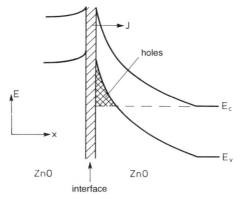

Fig. 3.31. Energy structure of the ZnO grain boundary at $V > V_{1\,\text{mA}}$ (reproduced from Mahan et al., 1979, by permission of the American Institute of Physics)

ponent. The voltage V_c differs from the varistor nominal voltage $V_{1\,\text{mA}}$ (usually $V_c < V_{1\,\text{mA}}$); it is defined as the voltage at which the current begins to increase rapidly, whereas $V_{1\,\text{mA}}$ is determined empirically. Mahan does not explain the mechanism of the creation of holes, but their presence have been confirmed by other investigators. Kim et al. (1991) suggest the occurrence of hole diffusion from the grain interior. Pike (1979, 1982) and Seager (1982) assume that the holes are created through ionization by collisions with the electrons that overcome the potential barrier. In Mahan's model, the J–V characteristic is described by the equation

$$J = J_0 S e^{E_b/kT} \Lambda(V) \qquad (3.24)$$

where $J_0 = \text{const}$, S is the normalized charge at the interface dependent upon the voltage, and $\Lambda(V)$ is the tunnelling factor equal to

$$\Lambda(V) = \frac{1}{u} \int_0^{E_b} dE\, e^{E/kT} e^{-W(E,V)} \qquad (3.25)$$

where

$$u = \int_0^{E_b} dE\, e^{E/kT} e^{-W(E,E_b)} \qquad (3.26)$$

$W(E, V)$ is the probability of a transmission through the potential barrier (which contributes to the value of the tunnelling factor). As the voltage increases and holes begin to be created (after the barrier becomes thinner), the factor $\Lambda(V)$ rapidly increases resulting in a very rapid increase of the current thereby increasing the nonlinearity of the J–V characteristic.

The atomic mechanism of the formation of energy structures at the electrically-active grain boundaries in ZnO varistors has not been fully explained yet. The investigators who attempted to explain this mechanism are Morris (1976) and Selim et al. (1980); they suggest that the potential barriers are formed thanks to the adsorption and diffusion of Bi ions.

Many authors have studied, by various techniques, the origin of the trap and surface states at ZnO interfaces. Using deep level transient spectroscopy (DLTS), Shohata et al. (1980, 1981) measured trap levels of energies of 0.24 and 0.33 eV. Examining General Electric commercial varistors by DLTS, Winston and Cordaro (1990) found deep interface traps of energy of 0.97 eV below the edge of the conduction band, with a density of states of 10^{15} m^{-2} and a capture cross-section of 4×10^{-19} m^{-2}. Rohatgi et al. (1988) studied a 0.26 eV trap level, which was identified to be a doubly charged zinc interstitial, $Zn_i^{\cdot\cdot}$. Nitayama et al. (1980) reported on energy levels of 0.18, 0.30 and 0.36 eV. Tanaka et al. (1990), using isothermal capacitance transient spectroscopy (ICTS), identified traps located 0.36 and 0.19 eV below the conduction band. They observed that after heat treatment in a slightly reduced atmosphere (90% N_2 + 10% O_2), the trap levels shifted by 0.16 and 0.09 eV, respectively, toward the conduction band. Complex-plane ac impedance spectroscopy (Macdonald, 1987) allows us to separate the bulk and grain boundary electrical properties and to calculate the position of trap states in an energy gap. For example, Abdullah et al. (1991) detected trap states of energies of 0.36, 0.06 and 0.01 eV below the edge of the conduction band. The ICTS and impedance spectroscopy examinations made by Maeda et al. (1989) revealed traps of energies of 0.60 and 0.27 eV. The ac impedance measurements can also be used to study varistor behaviour at elevated temperatures (Alim et al., 1988).

Einzinger (1982a,b) proposed an original model of the formation of potential barriers in ZnO–ZnO homojunctions which are formed when material is being cooled down from the sintering temperature. In the presence of oxide additives, the point defect equilibria in the near-surface region, including the equilibrium between the oxygen and zinc vacancies, are shifted. In addition, as a result of certain reactions between the individual phases, an epitaxial layer of stoichiometric ZnO forms on the grain surface (see Section 3.3). According to Einzinger's calculations, a sufficiently high potential barrier is then achieved. However, based on thermodynamic data given by Hagemark (1976), Mahan (1983) calculated that the Einzinger mechanism may only produce potential barriers with heights up to 0.1 eV. These discrepancies may result from possible inaccuracies in the thermodynamic data given by Hagemark.

In many papers published recently, the authors provide us with experimental evidence that oxygen plays the key role in establishing the nonlinear behaviour of ZnO varistors. Basic phenomena indicating how the interaction with oxygen is important here have long been known. It is known, for example, that sintering should be performed in an oxidizing atmosphere, and that by annealing in air or oxygen we can improve the varistor stability or restore the properties of an electrically degraded varistor.

Sukkar and Tuller (1988a,b) studied a ZnO–Ag model (a single crystal–evaporated silver or Mn-doped ZnO ceramics–Ag) to find how the presence of oxygen affects the J–V characteristic. The junctions examined rectified after treatment in an oxidizing atmosphere, and underwent degradation (which manifested itself by an increase of the leakage current) upon being exposed to a reducing atmosphere or to a current stress. The degradation was reversible and the junction recovered after its re-exposure to an oxidizing atmosphere. These effects suggest that chemisorbed oxygen plays an important role in the formation of the rectifying barrier.

Using an XPS/AES system, Stucki et al. (1987, 1990) examined the surface of a MOV varistor which was fractured in ultra high vacuum. Depth profiling combined with monitoring of Bi and O ions revealed that, in a degraded varistor, the grain boundaries were depleted of oxygen as compared with highly nonlinear samples. The quantity of the removed oxygen amounted to about one monolayer. Stucki et al. also confirmed the occurrence of the Bi segregation. They concluded that for the barrier to form it is necessary that a thin (0.5 nm) layer of bismuth should be adsorbed, and that its excess amount present at the grain boundary strongly affects the barrier height. Fujitsu et al. (1987, 1989) observed the barrier formation by oxygen chemisorbed on ZnO surfaces. These barriers were stabilized by Co addition to zinc oxide. The role of oxygen in the barrier formation was also suggested by Eda (1989) and by Raghu and Kutty (1992).

3.5 DEGRADATION OF THE ELECTRICAL PROPERTIES OF VARISTORS

The long-term operation under dc or ac voltages (lower than the varistor nominal voltage) and the short-term frequent overvoltage surges acting upon a varistor when it operates in an electric circuit may cause its degradation, which manifests itself in a change of the current–voltage characteristic (Fig. 3.32). As a result of degradation, the leakage current of the varistor increases, leading to an increased power dissipation and finally to the varistor destruction due to self-heating. It has been agreed that, to avoid this, the varistor voltage ($V_{1\,\mathrm{mA}}$) may only change by 10%. The working conditions of all varistor types are strictly specified.

Figure 3.32 indicates that the degree of degradation of a varistor may be described by comparing the voltages measured across it at the same

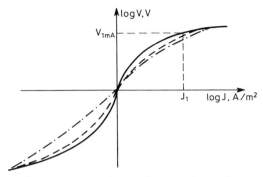

Fig. 3.32. Variation of the current-voltage characteristic of a ZnO varistor due to degradation: dashed line — by an ac current, dashed and dotted line — by a dc current or one-directional current pulses; at J_1, the current is 1 mA

current, or the currents measured at the same voltage, before and after degradation. The other conditions of the test (temperature, time, nature and magnitude of the load) should be strictly specified and kept constant. It should be noted that the changes of the current-voltage characteristic chiefly occur at current densities smaller than 10–100 mA/cm^2 (the leakage current region) and that, as the current density increases, these changes become increasingly smaller. The high-current portion of the current-voltage characteristic remains unchanged as the degradation proceeds.

Further in this chapter we shall discuss the phenomena that occur in metal-oxide varistors during degradation, the effect that degradation exerts upon the individual properties of these varistors, and the models proposed in the literature to explain the mechanism of degradation of their J–V characteristic.

3.5.1 Effect of degradation upon the J-V characteristic

As we can see from Fig. 3.32, the changes of the current-voltage characteristic due to the dc current are asymmetric: they are greater in that portion of the characteristic where the direction of the earlier degrading current was opposite to that of the present one. Pulse operation results in similar non-symmetric changes (Shirley and Paulson, 1979), whereas ac currents (50 Hz) alter the J–V characteristic in a symmetric manner. The degree of degradation increases as the applied load increases, irrespective of whether it is constant or pulsed. The degree of degradation thus increases with increasing voltages and with increasing time of operation. The rate of degradation (dV/dt or dJ/dt), however, decreases with time, settles after a certain time and, then, sometimes after several thousand hours, increases again. This last effect is a signal of a quick thermal runaway of the varistor (Sato, 1980).

An increase of temperature also activates the degradation process, greatly increasing its rate.

In varistors operating at ac voltages, the active current (i.e. the current other than that associated with the varistor capacitance) may not increase immediately after the voltage is applied but after a certain time known as the incubation period during which the current may even decrease. This effect is only observed in some specimens, especially at low temperatures (when the temperature increases, it vanishes).

At room temperature, after the degrading action is stopped, the current-voltage characteristic almost returns to its original shape. The mechanisms of this recovery are not, however, known exactly. Gupta et al. (1981; 1982a,b; 1983), for example, report that the time needed for the varistor to return to its original characteristic is much shorter than the time of degradation, whereas the varistors described by Bowen and Avella (1983) did not recover at all. Moldenhauer (1981) gives empirical formulae that describe the effects of temperature and time upon the degradation and recovery processes (cf. Subsection 3.5.5). At higher temperatures, the recovery processes proceed more rapidly. When, after the recovery, the degrading voltage is applied again, the varistor undergoes degradation anew: its behaviour is reversible. As already mentioned, the progressive degradation of the J–V characteristic of a varistor, especially at voltages greatly exceeding its admissible voltage, leads to increased power dissipation and, in consequence, to thermal breakdown of the varistor (Lee et al., 1977; Brückner et al., 1980b; Seitz et al., 1982).

3.5.2 Effect of degradation upon the dielectric properties of a varistor

Within the operating frequency range form 10^2 to 10^6 Hz, the capacitance of a varistor subjected to dc degradation decreases as the degradation progresses (Eda et al., 1980; Iga, 1980). At frequencies below 5×10^4 Hz, the dielectric loss coefficient, $\tan \delta$, increases with increasing degradation, but its changes may be restricted by applying an ac voltage. After the degrading agent is removed, the varistor capacitance slowly returns to its original value (the recovery effect).

In a dc-induced degradation, the minimum values of the varistor capacitance and of $\tan \delta$ (an equivalent, in a sense, of the varistor nominal voltage) in the diagram of Fig. 3.18a shift towards lower voltages as the degradation progresses. These voltage shifts are smaller when the polarizations of the current and preceding degrading agents are the same (Eda et al., 1980; Tomimuro and Terasaki, 1979).

On examining the variation of tan δ versus frequency in the range from 10^5 to 10^7 Hz we observe the occurrence of four dielectric absorption peaks that correspond to the energies 0.5, 0.3, 0.36 and 0.27 eV. As the degradation increases, these peaks change their positions (Hayashi et al., 1982).

3.5.3 Effect of heating upon the behaviour of degraded varistors

As already mentioned, when a degraded varistor is heated, it returns to its original behaviour. It has been found that the current-voltage characteristic of a degraded varistor may fully recover to its original shape after the varistor is heated at a temperature between 770 and 970 K. Eda et al. (1977, 1979c, 1980) and Shohata and Yoshida (1977) observed that when a dc voltage-degraded varistor was heated, a thermally stimulated current (the TSC) appeared, with a maximum occurring at a temperature between 620 and 820 K. The current was the greater, the higher the degrading voltage applied to the varistor and the longer the time during which the degradation had proceeded. The temperature at which the TSC reaches a maximum increases with increasing degradation time and temperature, but it is only slightly dependent on the degrading voltage (Eda and Matsuoka, 1977). The amplitude of the TSC and its variation with temperature was the same, irrespective of whether the varistor was heated in a nitrogen or an oxygen atmosphere. The recovery of the J–V characteristic of the varistor also occurs at the same time in both atmospheres, namely during the time when the TSC is observed (Eda et al., 1980).

When examining the pulse degradation of varistors, using current pulses of opposite polarization, Sato and Takada (1982) found that the TSC appears at both polarizations, just as observed in insulators. Charges of opposite signs are then accumulated, an effect which has not been observed after a long-term dc-current degradation.

3.5.4 Other effects associated with varistor degradation

Non-degraded varistors subjected to heating at a temperature between 770 and 970 K in air or in oxygen appear to be less liable to degradation. The heating, however, sometimes reduces the varistor nominal voltage (i.e., it increases the leakage current), but its shift may be decreased when the heating is carried out in oxygen atmosphere at a pressure of 1 MPa (Iga et al., 1976a). According to Brückner et al. (1980b) and Shohata and Yoshida (1977), the stability of the varistor may also be increased by adding a certain amount of glass frit (PbO–SiO_2–B_2O_3–ZnO) to the varistor material.

Examining the ZnO–Co_2O_3–Bi_2O_3 system with a high-resolution scan-

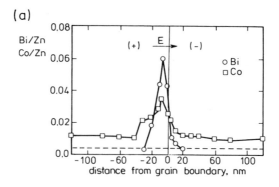

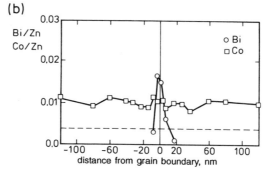

Fig. 3.33. Bi and Co concentration profiles at the boundaries of a ZnO grains in a degraded varistor: (a) at the boundary perpendicular to the direction of the electric field, (b) at the boundary parallel to the direction of the electric field; dashed line — approximate detection limit (reproduced from Chiang et al., 1982, by permission of the American Institute of Physics)

ning transmission electron microscope, Chiang et al. (1982) have found that, with increasing degradation, the Bi concentration profile changes at that ZnO grain boundary which is perpendicular to the direction of the electric field and which contains no intergranular phase. No such changes were observed at the boundary parallel to the electric field. Figure 3.33 shows the Bi and Co concentration profiles obtained by Chiang et al. From Fig. 3.33a we can see that in a degraded varistor the maximum Bi concentration shifts from the centre of the ZnO grain boundary to the interior of the grain (by a distance of about 5 nm) and that Co segregates at this boundary. These results have been the first to confirm the hypothesis of ion movements occurring in varistor ceramics during degradation.

Takahashi et al. (1982) report that the increase in the current that flows through a varistor at $V = $ const, when the varistor was heated at a temperature of 440 K in a nitrogen atmosphere (at a gas pressure of 0.4 MPa), was greater than that observed when the heating was carried out in an oxygen atmosphere under the same pressure (Fig. 3.34). This difference decreased

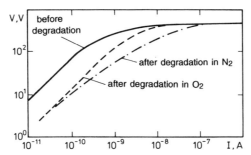

Fig. 3.34. Effect of an alternating electric field upon the current–voltage characteristic of a ZnO varistor at various atmospheres (reproduced from Takahashi et al., 1982, by permission of Elsevier Science Publishing Co., Inc.)

with decreasing temperature and vanished at 360 K. Within the gas pressure range from 0.1 to 0.6 MPa, the degree of degradation depended on the oxygen and nitrogen partial pressures only slightly. When the varistor was heated in air, the recovery of its J–V characteristic proceeded faster at temperatures between 330 and 420 K, but at 450 K it slowed down. When the heating was carried out in nitrogen, the J–V characteristic of the varistor did not recover at all.

Binešti et al. (1985, 1987) examined how the mass of a varistor varies when the varistor was heated at 920 K in oxygen whose partial pressure was $P_{O_2} = 10^{-21}$ MPa. They found that degraded varistors lost more mass than non-degraded ones, and that in the degraded varistors the reduction in mass was greater, and proceeded more quickly, in their outer portions as compared with their central portions, whereas in non-degraded varistors these effects were reversed. Binešti et al. attributed this reduction in varistor mass to the outflow of oxygen from the specimen. The values of the diffusion coefficients of oxygen in varistor ceramics, calculated from the results obtained by these investigators, were greater by about 6 orders of magnitude (e.g., 10^{-10} m$^2 \cdot$s^{-1}) than those measured in ZnO single crystals.

Einzinger (1978a) examined how the tendency to degradation of a varistor-type material depended on the degree of oxidation of manganese oxide dissolved in it. He found that the tendency to degradation decreased with increasing oxygen content in this oxide. The J–V characteristic was stabilized due to the oxidizing effect of H_2O_2. Einzinger (1981, 1982a) also examined the degradation of the J–V characteristic of the individual grain boundaries (microvaristors) present in varistor ceramics, and their C–V characteristic, where he found polarization occurring at low frequencies due to ion movements within the intergranular layer.

Eda (1982) suggested that varistors doped with Pr_6O_{11} in place of Bi_2O_3 are less liable to degradation.

3.5.5 Models of degradation mechanisms

A variety of models describing the mechanism of degradation of ZnO varistors based on the effects described above have been constructed.

The nonlinear conductivity of ZnO-based varistor ceramics is obtained by creating a specific energy structure of the boundaries of ZnO grains (see Section 3.4). From the fact that the changes in the J–V characteristic due to degradation occur in its low-current portion, where the behaviour of the varistor is determined by the parameters of the potential barriers present in it, we can infer that the degradation is caused by certain processes that affect these parameters. The degradation may be expected to occur as a result of a disturbance of the established equilibrium, since the potential barriers form as a result of the exchange of electric charges between the donor levels established in ZnO and the acceptor states, of the surface or trap type, gathered at the interface. The facts, however, that the degradation is a long-term process, that it is strongly activated thermally and that the thermally-stimulated current occurs during the course of recovery suggest that it is associated with ion movements within the material. A model of the degradation mechanism should thus answer the following questions:

— What factors determine the shape of the current-voltage characteristic when a given conduction mechanism is involved?

— The movements of what structure elements lead to a change of the current-voltage characteristic?

— What are the driving forces of these changes?

and it should

— explain all the experimental facts observed thus far, and

— describe the kinetics of the degradation process.

Since it is very difficult to fulfil all these requirements, none of the existing degradation models explain this phenomenon satisfactorily.

Eda et al. (1980) suggest that the degradation is due to the movements of ions within the intergranular phase and within the space charge layer formed inside the ZnO grain. These movements proceed in a direction perpendicular to the grain boundary (cf. Fig. 3.35). In a dc operation, anions and cations gather at the potential barriers formed on both sides of the intergranular phase. Ions from the intergranular phase approach one of these barriers, whilst those from the ZnO grain move towards the other barrier. Since the ions have different mobilities, the barriers are deformed in a nonsymmetric manner, which leads to degradation. Under ac conditions, the resultant path of the ions within the intergranular phase is zero (the effect is symmetric). Since, however, a strong internal field exists within the space charge region,

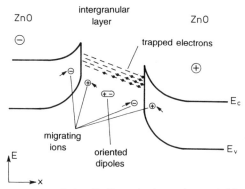

Fig. 3.35. Energy structure of the ZnO grain boundary at $V > 0$. Movements of the structure elements may cause varistor degradation (reproduced from Eda et al., 1980, by permission of the American Institute of Physics)

the resultant field is not zero and the ions shift symmetrically deforming the potential barriers. Eda et al. reject the hypothesis that oxygen ions take part in the degradation process, arguing that the thermally stimulated current (TSC) observed is the same irrespective of whether the process proceeds in an oxygen or in a nitrogen atmosphere. The basic disadvantage of this model lies in that it assumes the presence of an intergranular phase at the electrically-active ZnO grain boundary, whereas the examinations of the varistor microstructure have suggested that no intergranular phase occurs there.

In his later work, Eda (1982) takes into account this observation and suggests that the degradation process only proceeds at those boundaries where the intergranular phase occurs. He proposes an equivalent circuit of a varistor, from which he calculates that these individual effects may affect the characteristic of the varistor as a whole. This hypothesis is not, however, convincing in view of the fact, observed by Einzinger, that degradation also occurs in ZnO–ZnO homojunctions. Although Einzinger does not explain the mechanism of degradation, he suggests that this phenomenon is associated with ion movements and that it depends on the degree of oxidation of the ZnO boundaries.

Hayashi et al. (1982) have constructed their own model of the energy structure at the boundary of a ZnO grain, based on the measurements of the dielectric properties of varistors and of the TSC current. They suggest that the degradation proceeds in two stages: the first when the height and the width of the potential barrier are reduced, and the second during which its width alone continues to decrease. These effects are due to the movements of ions (perpendicularly to the grain boundary) within the space charge layer and to their gathering at the centre of the boundary.

Sato et al. (1980, 1982, 1983) constructed a completely different model of the degradation mechanism, which, according to them, is associated with the carrier trapping effect. They assume that certain electron trapping levels exist in the energy bandgap of the ZnO (inside the grains). During degradation, the concentration of the trapped electrons changes, resulting in the potential barriers on both sides of the grain boundary being deformed in a nonsymmetric manner. As to the two TSC peaks, opposite in sign, occurring after degradation, Sato et al. explain these in terms of an electron drift that take place in the two space charge layers in succession.

Moldenhauer (1981) and Bäther et al. (1983) give some numerical time and temperature data measured during the degradation and recovery processes. Like the other investigators mentioned earlier, they suggest that the degradation is caused by ion movements (not specified precisely) within the ZnO. These movements are induced by the electric field and lead to the formation of a chemical potential gradient that counteracts these movements. The relative change of the potential with time, at a constant current, is given by

$$\frac{\Delta U}{U \mathrm{d}t} = \frac{\mathrm{d}D}{\mathrm{d}t} = 2K(T)\sinh\left(\frac{eF}{2kT} - \frac{a'D}{kT}\right) \quad (3.27)$$

where $K(T) = g\lambda\omega_0 N_0 e^{-E/kT}$, in which λ is the average distance by which an ion jumps in the crystalline lattice, g the proportionality coefficient, ω_0 the frequency of ion jumps, N_0 the concentration of defects of various types, E the sum of enthalpy of the defect formation and the difference between the maximum and minimum ion enthalpy in the crystal lattice, $eF/2kT$ the change of the enthalpy E due to the action of the electric field F, $a'D/kT$ the change in the enthalpy E due to the chemical potential gradient, and a' a proportionality coefficient. In the derivation of this equation, the effects exerted by local temperature gradients and by the electron wind have been neglected.

When a varistor is in operation, the degradation and recovery processes proceed in it simultaneously (cf. equation (3.27)). The ions move perpendicularly to the ZnO grain boundary involved in conduction. The assumption that the activation energies of the degradation and of the recovery are equal to one another is however contrary to experiment. It follows from equation (3.27) that the recovery should proceed the more quickly the more advanced is the degradation. This statement has been contradicted by the fact, observed by Hozer (1988), that in some very strongly degraded varistors, recovery does not occur at all.

Gupta and Carlson (1985) describe another model of the degradation mechanism of the J–V characteristic in which they assume that the degra-

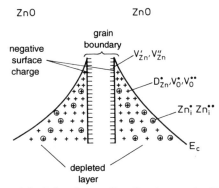

Fig. 3.36. Schematic model of the charge distribution at the ZnO grain boundary in a varistor-type material; V'_{Zn}, V''_{Zn} — acceptor-type Zn vacancies; $D^{\bullet}_{Zn}$ — trivalent ions (Bi or Sb) substituted into the Zn crystal lattice, $V^{\bullet}_O$, $V^{\bullet\bullet}_O$ — donor-type oxygen vacancies, $Zn^{\bullet}_i$, $Zn^{\bullet\bullet}_i$ — interstitial Zn ions (reproduced from Gupta and Carlson, 1985, by permission of Chapman & Hall)

dation is caused by movements of the interstitial Zn^+ ions present in the ZnO. The ions contained in the space charge layer are of the two kinds: trivalent Bi and Sb ions, which are immovable, and mobile interstitial Zn^+ ions. The charge of the donors is balanced by the vacancies V'_{Zn} and V''_{Zn} present on the interface (Fig. 3.36). When, after sintering, the varistor is cooled down, the near-boundary layer is almost entirely depleted of Zn^+ ions. Further away from the boundary, these ions are 'frozen' in the structure. The interstitial Zn ions, small in number, that remain near the grain boundary, can move relatively easily. It is they which cause degradation. Under the action of an electric field, especially at elevated temperatures, these ions move to the grain boundary and are deionized according to the equation

$$Zn^{\bullet}_i + V'_{Zn} = Zn^x_i + V^x_{Zn} \tag{3.28}$$

In this way, as a result of the neutralization of electric charge, the potential barriers are lowered. Under ac conditions, both the potential barriers degrade symmetrically. After the electric field is removed, the Zn_i atoms are ionized again and move back to the space charge layer. The stabilization achieved by heating the varistor at 870–1070 K is due to the removal of Zn^+_i ions from this layer. A quick oxygen diffusion then occurs along the ZnO grain boundaries, resulting in the reactions

$$O^x_O + V'_{Zn} = O'_O + V^x_{Zn} \tag{3.29}$$

A chemical potential gradient is formed and the ions diffuse towards the grain boundary, the more rapidly the higher the temperature. Then we have the reactions

$$V^x_{Zn} + Zn^{\bullet}_i = Zn^{\bullet}_{Zn} + V^x_i \tag{3.30}$$

$$Zn_{Zn}^{\cdot} + O_O' = ZnO \tag{3.31}$$

As donors are removed from the space charge layer and acceptors are removed from the interface, the potential barrier becomes lower and the nominal varistor voltage decreases. The temperature at which the varistor is heated should be optimized according to the composition of its material and its fabrication technology. The heating should remove all the Zn_i^+ ions present in the ceramics, but the temperature must not be so high as to produce and 'freeze' new Zn_i^+ ions in the space charge layer.

Thus far, Gupta and Carlson (1985) are the only investigators who describe the atomic degradation mechanism. However, like the models discussed earlier, their model is also controversial. Under the action of an ac sinusoidal field, the resultant electrodiffusion force that acts upon the ions is other than zero in the space charge region only. According to the model constructed by Gupta and Carlson, after degradation, the varistor characteristic returns to its original shape due to the ions moving away from the grain boundaries, in the direction opposite to that of the field produced by the potential barrier of considerable height. This assumption is contrary to the results obtained by the same investigators, which indicate that the activation energy of the recovery process is lower than the activation energy of degradation. Moreover, as already mentioned, recovery may even not occur at all.

In the opinion of the author of the present book, the degradation mechanism cannot be explained in terms of ion movements in the direction perpendicular to the electrically-active boundary of the ZnO grain, (i.e. in the direction parallel to the action of the electric field). Based on the observed relation, previously reported by Takahashi et al. (1982) and Binešti et al. (1985, 1987), among other authors, between the course of the degradation process and the action of oxygen, Hozer (1988, 1989) proposes another mechanism. By analysing the results of the so-called cyclic degradation test, during which the varistor is subjected to a dc degrading voltage applied cyclically and then left to recover at a predetermined temperature, it has been found that the degradation proceeds in two stages. After the first stage of degradation, the varistor recovers without any thermal activation, whereas after the second stage, the recovery requires an additional energy to be supplied to the varistor. Another observation was that in samples of cubic shape subjected to one-directional degradation, the current-voltage characteristic measured in the direction perpendicular to that of degradation had not been changed.

This effect may be interpreted as follows. At least part of the potential barrier forms due to the adsorption of oxygen ions on the interface when the material is being cooled down after sintering. The degrading action of the electric field causes these ions to desorb from the interface and to move along the ZnO grain boundaries towards the electrode. Due to the chemical potential gradients then established, the oxygen ions return to their previous positions after the voltage is removed, provided that they did not escape to the atmosphere. If they did, their return requires that a thermal activation energy be supplied.

The proposed model can explain most varistor properties observed experimentally, and is in good agreement with the general theory of the adsorption of oxygen on the ZnO surface (cf. Section 2.4). The relation between the properties of a varistor and the parameters of the sintering process, such as temperature and atmosphere, is associated with the establishment of appropriate equilibria of the oxygen that diffuses and is adsorbed on the ZnO grain surfaces. The different tendency of different varistors to degrade, depending on which Bi_2O_3 polymorph is present at the grain boundaries, can also be explained using this model. In contrast to the phase γ, the phase δ shows a high ionic conductivity for oxygen ions.

In previous sections we described some experimental evidence concerning the role of oxygen in the degradation of metal-oxide varistors. Its role has recently been strongly emphasized. Eda (1989) suggests that, during degradation, oxygen is transported along grain boundaries. Based on SMS observations, Sonder et al. (1985) found that, when varistors were heated above 560 K, oxygen became highly mobile. They suggest that macroscopic defects, such as cracks, can contribute substantially to the redistribution of oxygen and enhance the degradation of the varistor.

Philipp and Levinson (1983) consider, in general terms, the degradation of the J–V characteristics of metal-oxide varistors. They suggest that the electric field-induced degradation is only a particular case of the general degradation effects caused by various factors. The J–V characteristics change in a similar way under the action of such factors as:

— chemical factors (a modifications of the material composition),
— technological factors (preparation of the material),
— the kind of atmosphere in which the varistors are fabricated and operated,
— ambient pressure.

In view of the effects of a variety of factors, it is very difficult to estimate the service life of a varistor, based on the results of electrical measurements (such as the Arrhenius relationship) only. It is also evident that in the present state of knowledge it is impossible to formulate a universal mathematical description of the mechanism involved in the degradation of the J–V characteristics of metal-oxide varistors.

REFERENCES

Abdullah Al K., Bui A. and Loubiere A. (1991), Low-frequency and low-temperature behavior of ZnO-based varistor by ac impedance measurements, *J. Appl. Phys.*, **69** (7), 4046.

Aleksandrov V. V. and Pruzhinina V. I. (1979), Electrical properties of ZnO-based resistors RNS-60 (in Russian), *Elektrotekhnika*, **7**, 29.

Alim M. A., Seitz M. A. and Hirthe R. W. (1988), High-temperature/field alternating-current behavior of ZnO-based varistors, *J. Am. Ceram. Soc.*, **71** (1), C-52.

Alles A. B. and Burdick V. L. (1991), The effect of liquid-phase sintering on the properties of Pr_6O_{11}-based ZnO varistors, *J. Appl. Phys.*, **70** (11), 6883.

ASEA Catalogue (1982), Ed.1, File L, Part 2.

Asokan T., Ivengar G. N. K. and Nagabhushana G. R. (1987), Influence of process variables on microstructure and V–I characteristics of multicomponent ZnO-based nonlinear resistors, *J. Am. Ceram. Soc.*, **70** (9), 643.

Asokan T., and Freer R. (1990), Hot-pressing of zinc-oxide varistors, *Brit. Ceram. Trans. J.*, **89** (1), 8.

Avdeenko B. K., Bronfman A. I., Karachencev A. A., Kovalenko S. V., Makarova L. F., Potashev U. N., Shishman D. V. and Shchelokov A. I. (1976), ZnO-based ceramic resistors with a high nonlinearity coefficient (in Russian), *Elektrichestvo*, **9**, 61.

Avdeenko B. K. (1978), Electrical properties of ZnO ceramics produced at different annealing temperatures, *Izv. Akad. Nauk SSSR, Neorg. Mater.*, **14** (6), 1162.

Bäther K. H., Brückner W., Moldenhauer W. and Brückner H. P. (1983), Microscopic degradation model for ZnO varistors, *Phys. stat. solidi (a)*, **75**, (2), 465.

Baumgartner I. and Einzinger R. (1981), Epitaxial growth of ZnO-layers. A new aspect in homojunction models of ZnO varistors, in: *Sintering-Theory and Practice. Proc. 5th Intern. Round Table Conf. on Sintering*, Portoroz, Yug., 7-10 Sept.

Bernasconi J., Klein H. P., Knecht B. and Strassler S. (1976), Investigation of various models for metal oxide varistors, *J. Electron. Mater.*, **5** (5), 473.

Bernasconi J. (1977), Zinc oxide based varistors: a possible mechanism, *Solid State Commun.*, **21** (9), 867.

Bhushan B., Kashyap S. C. and Chopra K. L. (1981a), Novel nonohmic binary composite, *Appl. Phys. Lett.*, **38** (3), 160.

Bhushan B., Kashyap S. C. and Chopra K. L. (1981b), Electrical and dielectric behavior of a zinc oxide composite, *J. Appl. Phys.*, **52** (4), 2932.

Binešti D., Bonnet J. P., Onillon M. and Salmon R. (1985), Influence of the diffusion of oxygen on the aging of zinc oxide based varistors, *Int. Conf. Science of Ceramics 13*, Orleans, Sept. 9–11.

Binešti D. (1987), Aging and oxygen transfers in zinc oxide-based varistors, in: Vincenzini (ed.) (1987), *High-Tech Ceramics*, Elsevier, Amsterdam, 1801.

Bonasewicz P., Hirschwald W. and Neumann G. (1982), Diffusion of zinc and oxygen in non-stoichiometric zinc oxide, in: Nowotny J. (ed.) (1982), *Transport in Non-Stoichiometric Compounds*, PWN, Warszawa, Elsevier, Amsterdam, 153.

Bowen L. J. and Avella F. J. (1983), Microstructure, electrical properties and failure prediction in low clamping voltage zinc oxide varistors, *J. Appl. Phys.*, **54** (5), 2764.

Brückner W. (1980), Inhomogeneities and single barriers in ZnO-varistor ceramics, *Phys. stat. solidi (a)*, **59** (1), K1.

Brückner W., Moldenhauer W. and Hinz D. (1980), Thermal breakdown in ZnO-varistor ceramics, *Phys. stat. solidi (a)*, **59** (2), 713.

References

Carlson W. G. and Gupta T. K. (1982), Improved varistor nonlinearity via donor impurity doping, *J. Appl. Phys.*, **53** (8), 5746.

Castleberry D. E., (1979), Varistor-controlled liquid-crystal displays, *IEEE Trans. Electron. Dev.*, **ED-26** (8), 1123.

Chiang Y.-M., Kingery W. D. and Levinson L. M. (1982), Compositional changes adjacent to grain boundaries during electrical degradation of ZnO varistor, *J. Appl. Phys.*, **53** (3), 1765.

Clarke D. R. (1978), The microstructural location of the intergranular metal-oxide phase in a zinc oxide varistor, *J. Appl. Phys.*, **49** (4), 2407.

Clarke D. R. (1979), Grain-boundary segregation in a commercial ZnO-based varistor, *J. Appl. Phys.*, **50** (11), 6829.

Dereń J., Haber J. and Pampuch R. (1977), *Chemistry of Solid State* (in Polish), PWN, Warszawa.

Dorlanne O. and Tao M. (1987), The single grain junction in ZnO varistor, in: Vincenzini (ed.) (1987), *High-Tech Ceramics*, Elsevier, Amsterdam, 1809.

Driear J. M., Guertin J. P., Sokoly T. O. and Hackney L. B. (1981), Effect of dopant valence state on the microstructure of ZnO varistors, in: Levinson L. M., Hill D. C. (eds.) (1981), *Grain Boundary Phenomena in Electronic Ceramics. Advances in Ceramics*, Vol. 1, Am. Ceram. Soc., Columbus, Ohio, 316.

Eda K. and Matsuoka M. (1977), Thermally stimulated current in non-ohmic ZnO ceramics, *Jpn. J. Appl. Phys.*, **16** (1), 195.

Eda K. (1978), Conduction mechanism of non-ohmic zinc oxide ceramics, *J. Appl. Phys.*, **49** (5), 2964.

Eda K. (1979a), Transient conduction phenomena in non-ohmic zinc oxide ceramics, *J. Appl. Phys.*, **50** (6), 4436.

Eda K. and Matsuoka M. (1979b), Current oscillation phenomena in non-ohmic ZnO ceramics, *Jpn. J. Appl. Phys.*, **18** (5), 999.

Eda K., Iga A. and Matsuoka M. (1979c), Current creep in non-ohmic ZnO ceramics, *Jpn. J. Appl. Phys.*, **18** (5), 997.

Eda K., Iga A. and Matsuoka M. (1980), Degradation mechanism in non-ohmic zinc oxide ceramics, *J. Appl. Phys.*, **51** (5), 2678.

Eda K. (1982), Electrical properties of ZnO–Bi_2O_3 metal-oxide heterojunction—a clue of a role of intergranular layers in ZnO varistors, in: Pike G. E., Seager C. H. and Leamy H. J. (eds.) (1982), *Grain Boundaries in Semiconductors*, Elsevier, Amsterdam, 381.

Eda K. (1989), Zinc oxide varistors, *IEEE Electrical Insul. Mag.*, **5** (6), 28.

Einzinger R. (1975), Mikrokontakt Messungen am ZnO Varistoren, *Ber. Dtsch. Keram. Ges.*, **52** (7), 244.

Einzinger R. (1978a), Metal-oxide varistor action—a homojunction breakdown mechanism, *Appl. Surf. Sci.*, **1** (3), 329.

Einzinger R. (1978b), Dotierungseffekte in Metalloxid-Varistoren, *Ber. Dtsch. Keram. Ges.*, **55** (6), 329.

Einzinger R. (1979), Grain-junction properties of ZnO varistors, *Appl. Surf. Sci.*, **3** (3), 390.

Einzinger R. (1981), Grain boundary properties in ZnO varistors, in: Levinson L. M. and Hill D. C. (eds.) (1981), *Grain Boundary Phenomena in Electronic Ceramics. Advances in Ceramics*, Vol. 1, Am. Ceram. Soc., Columbus, Ohio, 359.

Einzinger R. (1982a), Nichtlineare Elektrische Leitfähigkeit von Dotierter Zinkoxid-Keramik, Doktors Thesis, Universität München.

Einzinger R. (1982b), Grain boundary phenomena in ZnO varistors, in: Pike G. E., Seager C. H. and Leamy H. J. (eds.) (1982), *Grain Boundaries in Semiconductors*, Elsevier, Amsterdam, 343.

Emtage P. R. (1977), The physics of zinc oxide varistors, *J. Appl. Phys.*, **48** (10), 4372.

Emtage P. R. (1979), Statistics and grain size in zinc oxide varistors, *J. Appl. Phys.*, **50** (11), 6833.

Fujitsu S., Toyoda H. and Yanagida H. (1987), Origin of ZnO varistor, *J. Am. Ceram. Soc.*, **70** (4), C-71.

Fujitsu S., Koumoto K. and Yanagida H. (1989), Formation of energy barrier by adsorbed oxygen on ZnO, *Solid State Ionics*, **32/33**, 482.

Gambino J. P., Philipp H. R., Pike G. E., Levinson L. M. and Kingery W. D. (1989), Effect of heat-treatments on the wetting behavior of bismuth-rich integranular phases in ZnO–Bi–Co varistors, *J. Am. Ceram. Soc.*, **72** (4), 642.

General Electric (1978), *Transient Voltage Suppression Manual*, sec. ed.

Graciet M., Salmon R., Flem G., Hagenmuller P., Hildebrandt M. and Buchy F. (1978), Évolution et rôle des constituants chimiques en cours du processus de fabrication des varistances a base d'oxyde de zinc, *Rev. de Physique Appliquée*, **13** (2), 67.

Graciet M. (1980), Physico-chemical properties and conduction mechanism in ZnO-based ceramic varistors, *Nouveau Journal de Chimie*, **4** (1), 29.

Graciet M. and Salmon R. (1981), Bismuth in metal oxide varistors. A new size of surge arresters, *The Bull. of the Bismuth Institute*, 3rd Quarter.

Greuter F. and Blatter G. (1990), Electrical properties of grain boundaries in polycrystalline compound semiconductors, *Semicond. Sci. Technol.*, **5** (2), 111.

Gupta T. K., Mathur M. P. and Carlson W. G. (1977), Effect of externally applied pressure on zinc oxide varistors, *J. Electron. Mater.*, **6** (5), 483.

Gupta T. K., Carlson W. G. and Hower P. L. (1981), Current instability phenomena in ZnO varistors under a continuous ac stress, *J. Appl. Phys.*, **52** (6), 4104.

Gupta T. K., Carlson W. G. and Hall B. O. (1982a), Metastable barrier voltage in ZnO varistors, in: Pike G. E., Seager C. H. and Leamy H. J. (eds.) (1982), *Grain Boundaries in Semiconductors*, Elsevier, Amsterdam, 393.

Gupta T. K. and Carlson W. G. (1982b), Barrier voltage and its effect on stability of ZnO varistor, *J. Appl. Phys.*, **53** (11), 7401.

Gupta T. K. and Carlson W. G. (1983), Defect induced degradation of barrier in ZnO varistor, in: Yan M. F. and Heuer A. H. (eds.) (1983), *Additives and Interfaces in Electronic Ceramics. Advances in Ceramics*, Vol. 7, Am. Ceram. Soc., Columbus, Ohio, 30.

Gupta T. K. and Carlson W. G. (1985), A grain boundary defect model for instability/stability of a ZnO varistor, *J. Mater. Sci.*, **20** (10), 3487.

Gupta T. K. (1990), Application of zinc-oxide varistors, *J. Am. Ceram. Soc.*, **73** (7), 1817.

Hagemark K. I. and Chacka L. C. (1975a), Electrical transport properties of Zn doped ZnO, *J. Solid State Chem.*, **15** (3), 261.

Hagemark K. I. and Toren P. E. (1975b), Determination of excess Zn in ZnO. The phase boundary Zn–ZnO, *J. Electrochem. Soc.*, **122** (7), 992.

Hagemark K. I. (1976), Defect structure of Zn doped ZnO, *J. Solid State Chem.*, **16** (3/4), 293.

Hampshire S. and Coolican J. (1987), Microstructural characterization of zinc oxide varistors in: Vincenzini (ed.) (1987), *High-Tech Ceramics*, Elsevier, Amsterdam, 1833.

Hannay N. B. (ed.) (1950), *Semiconductors*, Reinhold, New York.

References

Hardnen J. D., Martzloff F. D., Morris W. G. and Golden F. G. (1972), Metal oxide varistor: New way to suppress transients, *Electronics*, **45** (21), 91.

Harwig H. A. and Gerards A. G. (1978), Electrical properties of the α, β, γ and δ phases of bismuth sesquioxide, *J. Solid State Chem.*, **26** (3), 265.

Hayashi M., Haba M., Hirano S., Okamoto M. and Watanabe M. (1982), Degradation mechanism of zinc oxide varistors under dc bias, *J. Appl. Phys.*, **53** (8), 5754.

Hennings D. F. K., Hartung R. and Reijnen P. J. L. (1990), Grain-size control in low-voltage varistors, *J. Am. Ceram. Soc.*, **73** (3), 645.

Hieda S., Kobayashi M., Furuya N., Kondo N., Mitani K. and Aizawa T. (1975), Gapless lighting arresters for powder systems, *Meiden Rev.*, **46** (2), 17.

Hoffmann B. and Schwing U. (1981), Low voltage varistors, in: Levinson L. M. and Hill D. C. (eds.) (1981), *Grain Boundary Phenomena in Electronic Ceramics. Advances in Ceramics*, Vol. 1, Am. Ceram. Soc., Columbus, Ohio, 343.

Hohenberger G., Tomandl G., Ebert R. and Taube T. (1991), Inhomogeneous conductivity in varistor ceramics—Methods of investigation, *J. Am. Ceram. Soc.*, **74** (9), 2067.

Honnart F., Bolvin J. C., Thomas W. and de Vries K. J. (1983), Bismuth-lead oxide, a new highly conductive oxygen material, *Solid State Ionics*, **98** (10), 921.

Hower P. L. and Gupta T. K. (1979), A barrier model for ZnO varistors, *J. Appl. Phys.*, **50** (7), 4847.

Hozer L. and Szymański A. (1983), ZnO-based varistor material: microstructure and electrical properties (in Polish), *Prace ITME*, No. 9.

Hozer L. and Kołacz M. (1984), Effect of oxide additives on the nonlinearity coefficient of varistor-type cynkite ceramics (in Polish), *V Sympozjum Ceramiki*, Serock, Poland, Sept., 1984.

Hozer L. and Szymański A. (1984), Structural models for dielectric and conducting state of ZnO metal oxide varistor ceramics, *Applied Mineralogy*, *Proc. ICAM'84*, Feb. 22–25, Los Angeles, Calif.

Hozer L. and Kołacz M. (1985), Influence of oxide additives and sintering conditions on nonlinearity coefficient of varistor cynkite ceramics, *Proc. 9. Int. Baustoff und Silikattagung*, June 17–21, Weimar.

Hozer L. and Kozłowska H. (1987), Distribution of oxide additives in a varistor-type ZnO-based ceramics (in Polish), *Materiały Elektroniczne*, 1 (57), 28.

Hozer L. (1988), ZnO-based varistor-type ceramics: Degradation of its electrical properties under working conditions (in Polish), *Prace ITME*, No. 24.

Hozer L. (1989), Degradation of the J–V characteristic of ZnO varistors—a new mechanism (in Polish), *Materiały Elektroniczne*, 4 (64), 8.

Iga A., Matsuoka M. and Masuyama M. (1976a), Effect of heat-treatment on current creep phenomena in nonohmic zinc oxide ceramics, *Jpn. J. Appl. Phys.*, **15** (9), 1847.

Iga A., Matsuoka M. and Masuyama M. (1976a), Effect of phase transition of intergranular Bi_2O_3 layer in nonohmic ZnO ceramics, *Jpn. J. Appl. Phys.*, **15** (6), 1161.

Iga A. (1980), Drift phenomena of capacitance and current in nonohmic ZnO ceramics, *Jpn. J. Appl. Phys.*, **19** (1), 1201.

Inada M. (1978a), Crystal phases of nonohmic zinc oxide ceramics, *Jpn. J. Appl. Phys.*, **17** (1), 1.

Inada M. (1978b), Microstructure of nonohmic zinc oxide ceramics, *Jpn. J. Appl. Phys.*, **17** (4), 673.

Inada M. (1979), Effects of heat-treatment on crystal phases, microstructure and electrical properties of nonohmic zinc oxide ceramics, *Jpn. J. Appl. Phys.*, **18** (8), 1439.

Inada M. (1980), Formation mechanism of nonohmic zinc oxide ceramics, *Jpn. J. Appl. Phys.*, **19** (3), 409.

Inada M. and Matsuoka M. (1983), Formation mechanism of nonohmic ZnO ceramics in: Yan M.F. and Heuer A. H. (eds.) (1983), *Additives and Interfaces in Electronic Ceramics. Advances in Ceramics*, Vol. 7, Am. Ceram. Soc., Columbus, Ohio, 91.

Ivers-Tiffee E. and Seitz K. (1987), Characterization of varistor-type raw materials prepared by the evaporative decomposition of solutions technique, *Am. Ceram. Soc. Bull.*, **66** (9), 1384.

Jae Shi Choi and Chul Hyun Yo (1976), Study of the nonstoichiometric composition of zinc oxide, *J. Phys. Chem. Solids*, **37** (12), 1149.

van Kemenade J. T. C. and Ejinthoven R. K. (1975), Sintering of ZnO voltage dependent varistors, *Ber. Dtsch. Keram. Ges.*, **52** (7), 243.

van Kemenade J. T. C. and Ejinthoven R. K. (1978), Direct determination of barrier voltage in ZnO varistors, *Ber. Dtsch. Keram. Ges.*, **55** (6), 330.

van Kemenade J. T. C. and Ejinthoven R. K. (1979), Direct determination of barrier voltage in ZnO varistors, *J. Appl. Phys.*, **50** (2), 938.

Kim M. S., Oh H. H. and Kim C. K. (1991), Role of the hole diffusion current on the theory of conduction in ZnO varistor, *Jpn. J. Appl. Phys.*, **30** (11B), 1917.

Kingery W. D., van der Sande J. and Mitamura T. (1979), A scanning transmission electron microscopy investigation of grain-boundary segregation in $ZnO-Bi_2O_3$ varistor, *J. Am. Ceram. Soc.*, **62** (3–4), 221.

Klein H. P. and Strassler S. (1975), A simple model of a zinc oxide varistor, *Ber. Dtsch. Keram. Ges.*, **52** (7), 243.

Knecht B. and Klein H. P. (1978), Preparation and properties of ZnO varistors, *Ber. Dtsch. Keram. Ges.*, **55** (6), 326.

Kobayashi M. (1978), Development of zinc oxide nonlinear resistors and their applications to gapless surge arresters, *IEEE Trans. on Power Apparatus and Systems*, **PAS-97** (4), 1149.

Kosman M. and Petcold E. G. (1961), Fabrication of $ZnO-Bi_2O_3$ (in Russian), *Uchonye Zapiski LGPT im. A. Gercena*, **207**, 191.

Kostič P., Milosevič O., Zdravkovič Z. and Uskokovič D. (1987), Hot isostatic pressing of sintered ZnO varistor ceramics, in: Vincenzini (ed.) (1987), *High-Tech Ceramics*, Elsevier, Amsterdam, 1819.

Kröger F. A. and Vink H. J. (1956), Relations between the concentrations of imperfections in crystalline solids, in: Turnbull D. and Seitz F. (eds.) (1956), *Solid State Physics. Adv. Res. Appl.*, Vol. 3, 307.

Kröger F. A. (1973/74), *Chemistry of Imperfect Crystals*, North-Holland, Amsterdam.

Kutty T. R. N. and Raghu N. (1989), Varistors based on polycrystalline ZnO–Cu, *Appl. Phys. Lett.*, **54** (18), 1796.

Lauf R. J. and Bond W. D. (1984), Fabrication of high field zinc oxide varistors by sol-gel processing, *Am. Ceram. Soc. Bull.*, **63** (2), 278.

Lee J. J., O'Brien J. K. and Cooper M. S. (1977), Second-breakdown characteristics of metal-oxide varistors, *J. Appl. Phys.*, **48** (3), 1252.

Levin E. M. and Roth R. S. (1964), Polymorphism of bismuth sesquioxide. II. Effect of oxide additions on the polymorphism of Bi_2O_3, *J. Res. Nat. Bur. Stand. Physics and Chemistry*, **68A** (2), 197.

Levine D. (1975), Theory of varistor electronic properties, *CRC Crit. Rev. Solid State Sci.*, **5** (4), 597.

References

Levinson L. M. and Philipp H. R. (1975a), Conduction mechanism in metal oxide varistors, *J. Solid State Chem.*, **12** (3/4), 292.

Levinson L. M. and Philipp H. R. (1975b), The physics of metal oxide varistors, *J. Appl. Phys.*, **46** (3), 1332.

Levinson L. M. and Philipp H. R. (1976a), ac properties of metal oxide varistors, *J. Appl. Phys.*, **47** (3), 1117.

Levinson L. M. and Philipp H. R. (1976b), High-frequency and high-current studies of metal oxide varistors, *J. Appl. Phys.*, **47** (7), 3116.

Levinson L. M. and Philipp H. R. (1978a), Low temperature properties of metal oxide varistors, *J. Appl. Phys.*, **49** (12), 6142.

Levinson L. M. and Philipp H. R. (1978b), Behaviour of ZnO varistors at low temperatures, *Ber. Dtsch. Keram. Ges.*, **55** (8), 325.

Levinson L. M. and Hill D. C. (eds.) (1981), *Grain Boundary Phenomena in Electronic Ceramics. Advances in Ceramics*, Vol. 1, Am. Ceram. Soc., Columbus, Ohio.

Levinson L. M., Castleberry D. E. and Becker C. A. (1982), ZnO varistors for liquid crystal displays, *J. Appl. Phys.*, **53** (5), 3859.

Levinson L. M. and Philipp H. R. (1986), Zinc oxide varistors—a review, *Am. Ceram. Soc. Bull.*, **65** (4), 639.

Li P. W. and Hagemark K. I. (1975), Low temperature electrical properties of Zn doped ZnO, *J. Solid State Chem.*, **12** (3/4), 371.

Lou L. F. (1979), Current-voltage characteristics of $ZnO-Bi_2O_3$ heterojunction, *J. Appl. Phys.*, **50** (1), 555.

Lou L. F. (1980), Semiconducting properties of ZnO-grain-boundary-ZnO junctions in ceramic varistors, *Appl. Phys. Lett.*, **36** (7), 570.

Macdonald J. R. (ed.) (1987), *Impedance Spectroscopy*, John Wiley & Sons, New York.

Machine Design (1985), New varistors regulate very high voltages, **57** (10), 8.

Maeda T., Meguro S. and Takata M. (1989), Isothermal capacitance transient spectroscopy in ZnO varistor, *Jpn. J. Appl. Phys. Lett.*, **28** (4), L714.

Mahan G. D., Levinson L. M. and Philipp H. R. (1978), Single grain junction studies of ZnO varistor. Theory and experiment, *Appl. Phys. Lett.*, **33** (9), 830.

Mahan G. D., Levinson L. M. and Philipp H. R. (1979), Theory of conduction in ZnO varistors, *J. Appl. Phys.*, **50** (4), 2799.

Mahan G. D. (1982), Theory of ZnO varistors in: Pike G. E., Seager C. H. and Leamy H. J. (eds.) (1982), *Grain Boundaries in Semiconductors*, Elsevier, Amsterdam, 333.

Mahan G. D. (1983), Intrinsic defects in ZnO varistors, *J. Appl. Phys.*, **54** (7), 3825.

Mahan G. D. (1984), Fluctuations in Schottky barrier heights, *J. Appl. Phys.*, **55** (4), 980.

Matsuoka M., Masuyama T. and Iida Y. (1969), Voltage nonlinearity of zinc oxide ceramics doped with alkali earth metal oxide, *Jpn. J. Appl. Phys.*, **8** (10), 1275.

Matsuoka M. (1971), Nonohmic properties of zinc oxide ceramics, *Jpn. J. Appl. Phys.*, **10** (6), 736.

Matsuoka M. (1981), Progress in research and development of zinc oxide varistors, in: Levinson L. M. and Hill D. C. (eds.) (1981), *Grain Boundary Phenomena in Electronic Ceramics. Advances in Ceramics*, Vol. 1, Am. Ceram. Soc., Columbus, Ohio, 290.

Matsuura M. and Masuyama T. (1975), Negative resistance effect of a zinc oxide ceramics, *Jpn. J. Appl. Phys.*, **14** (6), 889.

Matsuura M. and Yamaoki H. (1977), Dielectric dispersion and equivalent circuit in nonohmic ZnO ceramics, *Jpn. J. Appl. Phys.*, **16** (7), 1261.

Medernach J. W. and Snyder R. L. (1978), Powder diffraction patterns and structures of the bismuth oxides, *J. Am. Ceram. Soc.*, **61** (11/12), 494.

Milosevič O., Kostič P., Petrovič V. and Uskokovič D. (1983), The study of the grain growth and electric properties of ZnO varistor ceramics, *Sci. Sintering*, **15** (3), 121.

Miyayama M., Terada H. and Yanagida H. (1981), Stabilization of β-Bi_2O_3 by Sb_2O_3 doping, *J. Am. Ceram. Soc.*, **64** (1), C19.

Miyayama M., Katsuta S., Suenaga Y. and Yanagida H. (1983a), Electrical conduction in β-Bi_2O_3 doped with Sb_2O_3, *J. Am. Ceram. Soc.*, **66** (8), 585.

Miyayama M., Suenaga Y. and Yanagida H. (1983b), Mixed electrical conduction in sintered bcc $6Bi_2O_3 \cdot SiO_2$, *J. Mater. Sci.*, **18** (10), 3023.

Miyoshi T., Maeda K., Takahashi K. and Yamazaki T. (1981), Effects of dopants on the characteristics of ZnO varistors, in: Levinson L. M. and Hill D. C. (eds.) (1981), *Grain Boundary Phenomena in Electronic Ceramics. Advances in Ceramics*, Vol. 1, Am. Ceram. Soc., Columbus, Ohio, 309.

Modine F. A. and Wheeler R. B. (1987), Fast pulse response of zinc oxide varistors *J. Appl. Phys.*, **61** (8), 3093.

Modine F. A., Wheeler R. B., Shim Y. and Cordaro J. F. (1989), Origin of the pulse response characteristics of zinc oxide varistors, *J. Appl. Phys.*, **66** (11), 5608.

Modine F. A., Major R. W., Choi S. I., Bergman L. B. and Silver M. N. (1990a), Polarization currents in varistors, *J. Appl. Phys.*, **68** (1), 339.

Modine F. A. and Wheeler R. B. (1990b), Pulse response characteristics of ZnO varistors, *J. Appl. Phys.*, **67** (10), 6561.

Mohanty G. P. and Azaroff L. V. (1961), Electron density distribution in ZnO crystals, *J. Chem. Phys.*, **35** (4), 1268.

Moldenhauer W. (1981), Degradation phenomena of ZnO varistors, *Phys. stat. solidi (a)*, **67** (2), 533.

Morris W. G. (1973), Electrical properties of ZnO–Bi_2O_3 ceramics, *J. Am. Ceram. Soc.*, **56** (7), 360.

Morris W. G. (1976), Physical properties of the electrical barriers in varistors, *J. Vac. Sci. Technol.*, **13** (4), 926.

Mukae K., Tsuda K. and Nagasawa I. (1977), Nonohmic properties of ZnO–rare earth–metal oxide–Co_3O_4 ceramics, *Jpn. J. Appl. Phys.*, **16** (8), 136.

Mukae K., Tsuda K. and Nagasawa I. (1979), Capacitance-vs-voltage characteristics of ZnO varistors, *J. Appl. Phys.*, **50** (6), 4475.

Mukae K. and Nagasawa I. (1981), Effect of praseodymium oxide and donor concentration in the grain boundary region of ZnO varistor, in: Levinson L. M. and Hill D. C. (eds.) (1981), *Grain Boundary Phenomena in Electronic Ceramics. Advances in Ceramics*, Vol. 1, Am. Ceram. Soc., Columbus, Ohio, 331.

Mukae K. (1987), Zinc oxide varistors with praseodymium oxide, *Am. Ceram. Soc. Bull.*, **66** (9), 1329.

Neumann G. (1981a), Non-stoichiometry and defect structure, in: Kaldis E. (ed.) (1981), *Current Topics in Materials Science*, Vol. 7 (*Zinc Oxide*), North-Holland, Amsterdam, 153.

Neumann G. (1981b), Diffusion and transport processes, in: Kaldis E. (ed.) (1981), *Current Topics in Materials Science*, Vol. 7 (*Zinc Oxide*), North-Holland, Amsterdam, 279.

Nitayama A., Sakaki H. and Ikoma T. (1980), Properties of deep levels in ZnO varistors and their effect on current-response characteristics, *Jpn. J. Appl. Phys.*, **19** (12), L743.

References

Okuma H., Amiji N., Suzuki M. and Tanno Y. (1983), Dielectric properties of Al-doped ZnO varistors, in: Yan M. F. and Heuer A. H. (eds.) (1983), *Additives and Interfaces in Electronic Ceramics. Advances in Ceramics*, Vol. 7, Am. Ceram. Soc., Columbus, Ohio, 41.

Olsson E. and Dunlop G. L. (1987), Development of intergranular microstructure in ZnO varistor materials, in: Vincenzini (ed.) (1987), *High-Tech Ceramics*, Elsevier, Amsterdam, 1765.

Olsson E., Dunlop G. L. and Osterlund R. (1989a), Development of interfacial microstructure during cooling of ZnO varistor material, *J. Appl. Phys.*, **66** (10), 5072.

Olsson E. and Dunlop G. L. (1989b), The effect of Bi_2O_3 content on the microstructure and electrical properties of ZnO varistor materials, *J. Appl. Phys.*, **66** (9), 4317.

Olsson E. and Dunlop G. L. (1989c), Characterization of individual interfacial barriers in a ZnO varistor material, *J. Appl. Phys.*, **66** (8), 3666.

Olsson E., Dunlop G. L. and Osterlund R. (1993), Development of functional microstructure during sintering of a ZnO varistor material, *J. Am. Ceram. Soc.*, **76** (1), 65.

Philipp H. R. and Levinson L. M. (1975), Tunneling of photoexcited carriers in metal oxide varistors, *J. Appl. Phys.*, **46** (7), 3206.

Philipp H. R. and Levinson L. M. (1976a), Optical method for determining the grain resistivity in ZnO-based ceramic varistors, *J. Appl. Phys.*, **47** (3), 1112.

Philipp H. R. and Levinson L. M. (1976b), Long-time polarization currents in metal-oxide varistors, *J. Appl. Phys.*, **47** (7), 3177.

Philipp H. R. and Levinson L. M. (1977), Low temperature studies on metal-oxide varistors. A clue to conduction mechanisms, *J. Appl. Phys.*, **48** (4), 1621.

Philipp H. R. and Levinson L. M. (1979), High-temperature behavior of ZnO-based ceramic varistors, *J. Appl. Phys.*, **50** (1), 383.

Philipp H. R. and Levinson L. M. (1981a), ZnO varistors for protection against nuclear electromagnetic pulses, *J. Appl. Phys.*, **52** (2), 1083.

Philipp H. R. and Levinson L. M. (1981b), Short-time pulse response of ZnO varistor grain boundaries, in: Levinson L. M. and Hill D. C. (eds.) (1981), *Grain Boundary Phenomena in Electronic Ceramics. Advances in Ceramics*, Vol. 1, Am. Ceram. Soc., Columbus, Ohio, 394.

Philipp H. R. and Levinson L. M. (1983), Degradation phenomena in zinc oxide varistors: A review, in: Yan M. F. and Heuer A. H. (eds.) (1983), *Additives and Interfaces in Electronic Ceramics. Advances in Ceramics*, Vol. 7, Am. Ceram. Soc., Columbus, Ohio, 1.

Philips Data Handbook (1984), Comp. and Mater., Pt 11.

Pike G. E. and Seager C. H. (1979), The dc voltage dependence of semiconductor grain-boundary resistance, *J. Appl. Phys.*, **50** (5), 3414.

Pike G. E. (1982), Electronic properties of ZnO varistors: New model, in: Pike G. E., Seager C. H. and Leamy H. J. (eds.) (1982), *Grain Boundaries in Semiconductors*, Elsevier, Amsterdam, 369.

Pike G. E., Seager C. H. and Leamy H. J. (eds.) (1982), *Grain Boundaries in Semiconductors*, Elsevier, Amsterdam.

Raghu N. and Kutty T. R. N. (1992), Relationship between nonlinear resistivity and the varistor forming mechanism in ZnO ceramics, *Appl. Phys. Lett.*, **60** (1), 100.

Rohatgi A., Pang S. K., Gupta T. K. and Straub W. D. (1988), The deep level transient spectroscopy studies of a ZnO varistor as a function of annealing, *J. Appl. Phys.*, **63** (11), 5375.

Safronov G. M., Batog V. N., Stepanyuk T. V. and Fedorov P. M. (1970), Bi_2O_3–ZnO equilibrium plot (in Russian), *Zhurnal Neorg. Chimii*, **16** (3), 865.

Sakshaug E. C., Kresge J. S. and Miske S. A. (Jr) (1977), A new concept in station arrester design, *IEEE Trans. Power Appar. Syst.*, **PAS-96** (2), 647.

Salmon R. (1980), Évolution de la caractéristique courant-tension des varistances à base d'oxyde de zinc avec la pression partielle d'oxygene de l'atmosphère de frittage, *Solid State Commun.*, **34** (5), 301.

Salmon R., Graciet M., Flem G. and Hagenmuller P. (1980), Influence des variétés allotropiques de Bi_2O_3 sur les caractéristiques électriques des varistances à base d'oxyde de zinc, *J. Solid State Chem.*, **34** (3), 377.

Samsonov G. V. (ed.) (1973), *The Oxide Handbook*, IFI/Plenum, New York.

Santhanam A. T., Gupta T. K. and Carlson W. G. (1979), Microstructural evaluation of multicomponent ZnO ceramics, *J. Appl. Phys.*, **50** (2), 852.

Sato K. (1980), Electrical conduction in ZnO varistors under continuous dc stress, *Jpn. J. Appl. Phys.*, **19** (5), 909.

Sato K. and Takada Y. (1982), A mechanism of degradation in leakage currents through ZnO varistors, *J. Appl. Phys.*, **53** (12), 8819.

Sato K., Takada Y., Takamura T. and Ototake M. (1983), Carrier trapping model of degradation in ZnO varistors, in: Yan M. F. and Heuer A. H. (eds.) (1983), *Additives and Interfaces in Electronic Ceramics. Advances in Ceramics*, Vol. 7, Am. Ceram. Soc., Columbus, Ohio, 22.

Schwing U. (1980), ZnO single crystals with an intermediate layer of metal oxides—a macroscopic varistor model, *J. Appl. Phys.*, **51** (8), 4558.

Schwing U. and Hoffmann B. (1981), New approach to the measurement of the single-contact varistor, in: Levinson L. M. and Hill D. C. (eds.) (1981), *Grain Boundary Phenomena in Electronic Ceramics. Advances in Ceramics*, Vol. 1, Am. Ceram. Soc., Columbus, Ohio, 383.

Schwing U. (1984), Ein Experimentalles Modell zur Beschreibung des Varistorkontaktes in Dotierter Zinkoxidker Keramik, Doktors Thesis, Karlsruhe.

Seager C. H. (1982), The electronic properties of semiconductor grain boundaries, in: Pike G. E., Seager C. H. and Leamy H. J. (eds.) (1982), *Grain Boundaries in Semiconductors*, Elsevier, Amsterdam, 85.

Seitz M. A., Hirthe R. W. and Potter M. E. (1982), Thermal runaway in metal oxide varistors, in: Pike G. E., Seager C. H. and Leamy H. J. (eds.) (1982), *Grain Boundaries in Semiconductors*, Elsevier, Amsterdam, 405.

Seitz M. A., Ivers-Tiffee E., Thomann H. and Weiss A. (1987), Influence of zinc acetate and nitrate salts on the characteristics of undoped ZnO powders, in: Vincenzini (ed.) (1987), *High-Tech Ceramics*, Elsevier, Amsterdam, 1753.

Selim F. A., Gupta T. K., Hower L. and Carlson W. G. (1980), Low voltage ZnO varistor. Device process and defect model, *J. Appl. Phys.*, **51** (1), 765.

Shirley C. G. and Paulson W. M. (1979), The pulse degradation characteristic of ZnO varistors, *J. Appl. Phys.*, **50** (9), 5782.

Shohata N. and Yoshida J. (1977), Effect of glass on nonohmic properties of ZnO ceramic varistors, *Jpn. J. Appl. Phys.*, **16** (12), 2299.

Shohata N., Matsumura T. and Ohno T. (1980), DLTS measurement on nonohmic zinc oxide ceramic varistor, *Jpn. J. Appl. Phys.*, **19** (9), 1793.

Shohata N., Matsumura T., Utsumi K. and Ohno T. (1981), Properties of multilayer ZnO ceramic varistors, in: Levinson L. M. and Hill D. C. (eds.) (1981), *Grain Boundary*

Phenomena in Electronic Ceramics. Advances in Ceramics, Vol. 1, Am. Ceram. Soc., Columbus, Ohio, 349.

Siemens (1978/79), *SIOV Metalloxid Varistoren*—catalogue.

Siemens (1980/81), *SIOV Metalloxid Varistoren*—catalogue.

Snow G. S. and Cooper R. A. (1980), Hot press with flat plate heater and its application to the fabrication of large varistor slugs, *Am. Ceram. Soc. Bull.*, **59** (5), 550.

Snow G. S., White S. S., Cooper R. A. and Armijo J. R. (1980), Characterization of high field varistors in system ZnO–CoO–PbO–Bi_2O_3, *Am. Ceram. Soc. Bull.*, **59** (6), 617.

Sonder E., Levinson L. M. and Katz W. (1985), Role of short-circuiting pathways in reduced ZnO varistors, *J. Appl. Phys.*, **58** (11), 4420.

Stanisic G. (1984), Contribution to better understanding of sintering mechanism of ZnO–Bi_2O_3 system, *Sci. Sintering*, **16** (2), 121.

Stevenson D. A. (1973), Diffusion in the chalcogenides of Zn, Cd and Pb, in: Shaw D. (ed.) (1973), *Atomic Diffusion in Semiconductors*, Plenum Press, London.

Stucki F., Bruesch P. and Greuter F. (1987), Electron spectroscopic studies of electrically active grain boundaries in ZnO, *Surf. Sci.*, **189/190**, 294.

Stucki F. and Greuter F. (1990), Key role of oxygen at zinc-oxide varistor grain-boundaries, *Appl. Phys. Lett.*, **57** (5), 446.

Sukkar M. H. and Tuller H.L. (1983), Defect equilibria in ZnO varistor materials, in: Yan M. F. and Heuer A. H. (eds.) (1983), *Additives and Interfaces in Electronic Ceramics. Advances in Ceramics*, Vol. 7, Am. Ceram. Soc., Columbus, Ohio, 71.

Sukkar M. H. and Tuller H.L. (1988a), ZnO interface electrical properties—role of oxygen chemisorption, in Nowotny J. and Weppner W. (eds.), *Non-Stoichiometric Compounds, Surfaces, Grain Boundaries and Structural Defects*, Kluwer Academic Publ., 237.

Sukkar M. H. and Tuller H.L. (1988b), Electrical characteristics of single ZnO grain boundaries, in: Tuller H. L. and Smyth D. M. (eds.) *Electro-Ceramics and Solid-State Ionics, Proceedings*, Vol. 88-3, Electrochemical Society, Inc., 61.

Takahashi T. and Iwahara H. (1973), High oxide ion conduction in the sintered oxides of the system Bi_2O_3–WO, *J. Appl. Electrochem.*, **3** (1), 65.

Takahashi T., Esaka T. and Iwahara H. (1975a), High oxide ion conduction in the sintered oxides of the system Bi_2O_3–Gd_2O_3, *J. Appl. Electrochem.*, **5** (3), 197.

Takahashi T., Iwahara H. and Arao T. (1975b), High oxide ion conduction in the sintered oxides of the system Bi_2O_3–Y_2O_3, *J. Appl. Electrochem.*, **5** (3), 187.

Takahashi T., Iwahara H. and Esaka T. (1977), High oxide ion conduction in the sintered oxides of the system Bi_2O_3–M_2O_5, *J. Electrochem. Soc.*, **124** (10), 1563.

Takahashi T. and Iwahara H. (1978), Oxide ion conductors based on bismuth sesquioxide, *Mat. Res. Bull.*, **13** (12), 1447.

Takahashi T., Miyoshi T., Maeda K. and Yamazaki T. (1982), Degradation of zinc oxide varistors, in: Pike G. E., Seager C. H. and Leamy H. J. (eds.) (1982), *Grain Boundaries in Semiconductors*, Elsevier, Amsterdam, 399.

Takemura T., Kobayashi M., Takada Y. and Sato K. (1983), Effects of aluminium as dopant on the characteristics of ZnO varistors, in: Yan M. F. and Heuer A. H. (eds.) (1983), *Additives and Interfaces in Electronic Ceramics. Advances in Ceramics*, Vol. 7, Am. Ceram. Soc., Columbus, Ohio, 50.

Takemura T., Kobayashi M., Takada Y. and Sato K. (1986), Effects of bismuth sesquioxide on the characteristics of ZnO varistors, *J. Am. Ceram. Soc.*, **69** (5), 430.

Takemura T., Kobayashi M., Takada Y. and Sato K. (1987), Effects of antimony oxide on the characteristics of ZnO varistors, *J. Am. Ceram. Soc.*, **70** (4), 237.

Tanaka J., Hishita S. and Okushi H. (1990), Deep levels near the grain-boundary in a zinc oxide varistor—energy change due to electrical degradation, *J. Am. Ceram. Soc.*, **73** (5), 1425.

Tao M., Bui Ai, Dorlanne O. and Loubiere A. (1987), Different single grain junctions within a ZnO varistor, *J. Appl. Phys.*, **61** (4), 1562.

Tomimuro H. and Terasaki Y. (1979), Degradation mechanism of ZnO varistors, *Jpn. J. Appl. Phys.*, **18** (8), 1653.

Tominaga S., Azumi K., Nitta T., Nagai N., Imataki M. and Kuwabara H. (1979a), Reliability and application of metal oxide surge arresters, *IEEE Trans. on Power Appar. Syst*, **PAS-98** (3), 805.

Tominaga S., Azumi K., Shibuya Y., Imataki M., Fujiwara Y. and Nishida S. (1979b), Protective performance of metal oxide surge arrester based on the dynamic $V-I$ characteristics, *IEEE Trans. on Power Appar. Syst*, **PAS-98** (6), 1860.

Tominaga S. (1980), Stability and long-term degradation of metal oxide surge arresters, *IEEE Trans. on Power Appar. Syst*, **PAS-99** (4), 1548.

Trontelj M., Kolar D. and Kraševec V. (1983), Influence of additives on varistor microstructure, in: Yan M. F. and Heuer A. H. (eds.) (1983), *Additives and Interfaces in Electronic Ceramics. Advances in Ceramics*, Vol. 7, Am. Ceram. Soc., Columbus, Ohio, 107.

Tsuda K. and Mukae K. (1987), Interface states of zinc oxide varistors, in: Vincenzini (ed.) (1987), *High-Tech Ceramics*, Elsevier, Amsterdam, 1781.

Vandanamme L. K. J. and Brugman J. C. (1980), Conduction mechanism in ZnO varistors, *J. Appl. Phys.*, **51** (8), 4240.

Vincenzini P. (ed.) (1987), *High-Tech Ceramics*, Elsevier, Amsterdam.

Williams P., Krivanek O. L. and Thomas G. (1980), Microstructure-property relationships of rare-earth-zinc oxide varistors, *J. Appl. Phys.*, **51** (7), 3930.

Winston R. A. and Cordaro J. F. (1990), Grain-boundary interface electron traps in commercial zinc-oxide varistors, *J. Appl. Phys.*, **68** (12), 6495.

Wong J. (1974), Nature of intergranular phase in nonohmic ZnO ceramics containing 0.5 mol% Bi_2O_3, *J. Am. Ceram. Soc.*, **57** (8), 357.

Wong J. (1975), Microstructure and phase transformation in a highly nonohmic metal oxide varistor ceramics, *J. Appl. Phys.*, **46** (4), 1653.

Wong J., Rao P. and Koch E. F. (1975), Nature of an intergranular thin film phase in a highly nonohmic metal oxide varistor, *J. Appl. Phys.*, **46** (4), 1827.

Wong J. (1976), Barrier voltage measurement in metal oxide varistors, *J. Appl. Phys.*, **47** (11), 4971.

Wong J. (1980), Sintering and varistor characteristics of $ZnO-Bi_2O_3$ ceramics, *J. Appl. Phys.*, **51** (8), 4453.

Yan M. F. and Heuer A. H. (eds.) (1983), *Additives and Interfaces in Electronic Ceramics. Advances in Ceramics*, Vol. 7, Am. Ceram. Soc., Columbus, Ohio.

CHAPTER 4

Positive Temperature Coefficient of Resistivity (PTCR) Thermistors

Since the mid 1950s, growing attention has been paid to barium titanate ($BaTiO_3$)-based materials which exhibit a positive temperature coefficient of resistivity. Just as with the group of ZnO-based materials, the properties of the large group of $BaTiO_3$-based materials are determined by the potential barriers formed at the grain boundaries of the basic $BaTiO_3$ phase. The first investigators to report on PTCR materials, discovered in 1950, were Haayman et al. (1955). Since that time, the production of these materials has been growing very quickly. How it developed during the years has been described by Kulwicki (1981); according to his estimates the production volume amounted to 10^8 devices per year.

An example of the variation of the resistance of the PTCR materials with temperature is shown in Fig. 4.1. The PTCR materials are used within a relatively narrow temperature range, from 248 to 428 K and, at the maximum voltage, from 273 to 328 K (Philips, 1984). The basic parameters given in catalogues also include the values of the resistance at various temperatures, e.g., at 298 and 353 K, the temperature at which the resistance begins to increase (when the resistance of a given device is doubled), and the maximum value of the coefficient representing the variation of the resistance with temperature (in %/K). The catalogues also give the temperature-time constant characterizing the time necessary for the device to change its temperature under given conditions (this time is of the order of several to several tens of seconds), and certain important voltage values, such as the voltage that may be applied to the device at a temperature of 328 K. The power dissipation factor and the heat capacity of the device are also given. The catalogues also specify other parameters important from the point of view of

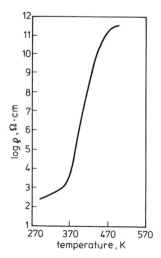

Fig. 4.1. Example of the resistance-temperature variation in a PTCR thermistor (the resistivity jump may exceed 7 orders of magnitude)

current application of the device and the characteristics of the device such as the temperature-resistance characteristic, the current-voltage characteristic, and the variation of the resistance with temperature on cooling.

The variation of the resistance with temperature, which in the earliest PTCR materials did not exceed three orders of magnitude (Saburi, 1961), in modern materials may even amount to eight orders of magnitude when the temperature changes by several tens of degrees. This has been achieved thanks to the development of an understanding of the fundamental physical phenomena that underlie this effect; the technological regimes developed based on this knowledge must however be strictly observed if we wish to produce a device with the high positive temperature coefficient of resistivity. Just as in varistor type materials, the structure and properties of the boundaries of the semiconducting $BaTiO_3$ grains are the least understood areas.

The PTCR effect occurs as a result of the ferroelectric–paraelectric phase transformation that takes place in sintered $BaTiO_3$ doped with other metals, such as Sb and Y, with rare earth metals, such as La, Er and Ho (located at the Ba sites), or with Ta and Nb located at the Ti lattice sites (Daniels et al., 1978/1979). These metals form donor centres in the $BaTiO_3$. In practice, pure $BaTiO_3$ is often replaced by its solid solutions with strontium and lead titanates, such as $(Ba_{1-x}Sr_x)TiO_3$ or $(Ba_{1-x}Pb_x)TiO_3$.

The PTCR effect may be enhanced by adding small amounts of certain substances, such as oxides of Mn, Cr, Cu, Fe, V and Ru, which introduce deep acceptor levels, thereby increasing the changes of resistivity with temperature by 6 to 7 orders of magnitude within narrow temperature range.

The PTCR ceramics often also contains certain amounts of Al_2O_3, SiO_2 and some excess of TiO_2, which after the eutectic reaction with $BaTiO_3$ occurs during sintering, form a liquid phase. Some of these additives, such as Al_2O_3 and SiO_2, may be introduced in an uncontrolled way during the fabrication process (e.g., milling contaminations). In order to obtain the required structure of the grain boundaries and of the interior of the $BaTiO_3$ grains, it is very important that the heat treatment of the material is carried out in an appropriate atmosphere.

Initially, PTCR thermistors were used for detecting temperature variations or the occurrence of the melting point of a material. Very soon they found application as self-controlled heating devices. The first large-scale application of these thermistors was as TV degaussers introduced in 1968; here they worked as the switch-on overvoltage current limiters. They are still used for this purpose in millions of modern colour TV sets manufactured today. The PTCR materials are also used for fabricating constant-temperature heating devices in which the temperature is stabilized due to the equilibrium heat exchange that occurs between varying-resistivity devices, operated at a constant voltage, and the surroundings (Heywang et al., 1976). Such devices are, for example, used as fuel evaporators, electric current limiters, motor starters, fuel heaters intended for freezers and air-conditioning equipment, sequential time switches for turning on electric heaters, overheating protectors for various electric devices, electric current stabilizers and alarm devices. In the field of home appliances, these PTCR heating elements are used in hair dryers, refrigerators, irons, food warming plates, etc. The Philips catalogue (1984) divides the applications of PTCR thermistors into two basic classes:

(1) applications where the temperature of the thermistor is primarily determined by the temperature of the ambient medium,

(2) applications where the temperature of the thermistor is primarily determined by the current flowing through it.

Various designs are proposed for a variety of applications. In this book, however, we shall refer to the basic design version, which is similar to that of a metal-oxide varistor (i.e., a ceramic pellet with the electrodes deposited on its faces).

Inaba et al. (1992) propose using PTCR thermistors as infrared detecting elements. Because of the large variation of their resistivity with temperature, PTCR thermistors are expected to have a higher sensitivity than widely used NTC thermistors. Their operating temperature must, however, be maintained strictly in the vicinity of the Curie point.

Later in the book we shall discuss in detail the electrical properties and structure of PTCR materials. The discussion will be focused on the mechanism of the occurrence of the PTCR effect. First, however, we briefly remind the reader the properties of barium titanate, which is the major constituent of these materials.

4.1 PROPERTIES OF BaTiO$_3$

At temperatures above the Curie point (about 283 K), barium titanate is a paraelectric material of the perovskite structure showing cubic symmetry. At the Curie point (on cooling), the paraelectric is transformed spontaneously into a ferroelectric, the structure becomes polarized and certain shifts (by about 1% of the lattice constant) occur in the crystal lattice. In effect, the structure becomes tetragonal, as shown in Fig. 4.2.

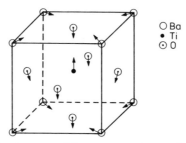

Fig. 4.2. A perovskite lattice type. The arrows indicate the shifts of the atoms that result in the hexagonal structure transforming into tetragonal (after Haywang, 1971; reprinted by permission of Chapman & Hall)

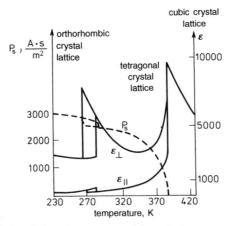

Fig. 4.3. Changes of the dielectric constant (ε) and the spontaneous polarization (P_s) due to the phase transformations that occur in a barium titanate crystal; $\varepsilon_\parallel$ and $\varepsilon_\perp$ are the dielectric constants measured in parallel and perpendicularly, respectively, to the polarization vector (after Haywang, 1971; reprinted by permission of Chapman & Hall)

These changes in the lattice structure and the occurrence of the spontaneous polarization are accompanied by a change in the dielectric constant of the BaTiO$_3$ (see Fig. 4.3); the cubic-into-tetragonal transformation results in the dielectric constant becoming anisotropic. As the temperature decreases, the value of this constant also decreases and, then, $\varepsilon_\perp$ (perpendicular to the polarization) begins to increase again. The crystal lattice becomes polarized along all directions; at 278 K the structure becomes orthorhombic and, then, at about 200 K (not shown in the figure), rhombohedral. In this way the ferroelectric-like domain structure is being formed.

Above the Curie point, the changes of the dielectric constant are described by the Curie–Weiss law

$$\varepsilon = \frac{C}{T - T_C} \qquad (4.1)$$

where C is the Curie constant, and T_C the Curie temperature. We can see that, after the Curie temperature is exceeded, the value of the dielectric constant rapidly decreases.

The width of the BaTiO$_3$ bandgap is about 3 eV (Casella and Keller, 1959). When undoped, this material behaves as an insulator; at room temperature its resistivity is of the order of 10^{10} $\Omega \cdot$cm. It is, however, relatively easy to obtain the n-type conduction by, for example, reducing the Ti^{+4} ions to Ti^{+3} ions. This reduction may be effected by heating the material in a reducing atmosphere, which produces oxygen vacancies, or by substituting (through appropriate doping) Ba^{+2} or Ti^{+4} ions with ions of higher valencies. Gersthen et al. (1965a, b) report on small-polaron type conduction in barium titanate. The ways in which BaTiO$_3$ may be doped are discussed in further chapters of the book.

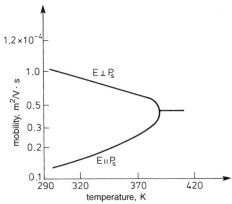

Fig. 4.4. Anisotropy of the electron mobilities in a BaTiO$_3$ crystal varies with temperature (below the Curie point): E is the electric field (after Berglund and Baer, 1967)

As with ZnO, it is difficult to obtain p-type $BaTiO_3$ at room temperature. This cannot be achieved even if the material is doped with ions of lower valencies. The defect structure of barium titanate has been extensively studied in recent years and the interested reader is referred to the literature (Daniels et al., 1978/1979; Desu and Payne, 1990a, b, c, d; Chiang and Takagi, 1990a, b, 1992).

Above the Curie point, the electron mobility is about 0.5 $cm^2/V \cdot s$ and, below this point, it becomes anisotropic (Berglund and Baer, 1967). Its variation with temperature is shown in Fig. 4.4.

4.2 PROCESS ENGINEERING AND MICROSTRUCTURE OF PTCR MATERIALS

Although the fabrication of PTCR materials may be considered as standard ceramic technology, it should be realized that it requires particular care in controlling the individual process parameters. According to Ueoka and Yodogawa (1974), these parameters include:

(1) high purity of starting materials;
(2) precise control of the dopant content;
(3) high cleanliness of the process;
(4) control of the physical and chemical properties of the powders;
(5) precise control of the sintering process.

As will be shown later, the doping level, the impurity content and the sintering process parameters have a decisive effect upon the properties of the final material.

The first stage of the process involves mixing (often using a ball mill with SiO_2 or agate balls) the powders in the specified proportions. The starting powders are rarely oxides, they are more often carbonates (of Ba or Sr), nitrates (of La) or even sulphates (of Mn). The TiO_2 content is usually somewhat greater than stoichiometric (about 0.5 to 1.0 mol% above the stoichiometric proportion) as this makes the sintering process easier to perform and improves the properties of the final product.

After the material is mixed in water or alcohol and dried, some plasticizers are added and pellets are pressed out. The pellets are next sintered, usually in air, at a temperature of about 1670 K and then cooled down to room temperature in a controlled manner. The density of the ceramics thus obtained of course depends on its composition and usually is 5.7 to 5.8 g/cm^3. The process often also involves calcination at a temperature of, for example, 1370 K, and remixing.

The powder mixture may also be prepared using the sol-gel technique (Chaput and Boilot, 1987). Phule et al. (1987) and Young et al. (1987)

report that they have used organometallic compounds as the starting materials. Kuwabara and Yanagida (1973) introduced Bi_2O_3 by immersing the specimen, after it had been partially sintered, in an alcohol solution of bismuth chloride and, then, continuing the sintering process.

The ideal microstructure of PTCR materials is simple: they should be single-phase materials with well defined uniform $BaTiO_3$ grains that contain dissolved donor impurities. The grain size ranges from a few to several tens of micrometres and the grain size distribution should be narrow. The additives that affect the surfaces of the barium titanate grains should not form precipitates at the boundaries and in the interior of the grains.

The available literature on the subject gives numerous values of the typical parameters of the ceramic materials, such as the density, porosity and grain size, obtained at different process parameters. In view of the great variety of these data, it is impossible to quote all of them here. By way of example, we only mention that Yoneda et al. (1976) describe how the density, resistivity and average grain diameter of the $BaTiO_3$ grains vary as functions of the sintering temperature over the range from 1470 to 1610 K for $Ba_{0.776}Sr_{0.22}Y_{0.004}TiO_3$ doped with 0.75 wt% SiO_2 and 0.042 wt% $MnCO_3$ (Fig. 4.5).

When the doping level is increased, in order for example to increase the conductivity of the $BaTiO_3$ grains, the growth of the grains is retarded.

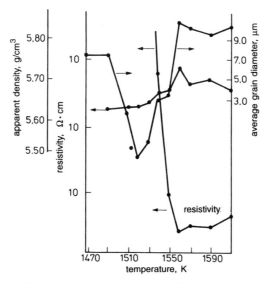

Fig. 4.5. Variations of the apparent density, resistivity and average grain diameter of the $BaTiO_3$ grains in a PTCR material as functions of the sintering temperature (after Yoneda et al., 1976; reprinted by permission of the American Ceramic Society)

Above a certain critical dopant content (of the order of 0.3 mol%), the resistivity of the material rapidly increases, since the growth of the grains is almost entirely inhibited. This effect, known as the *doping anomaly*, will be discussed in Subsection 4.4.3. Janitzki et al. (1979) reported that, in undoped $BaTiO_3$, they easily obtained an average grain diameter of the order of 100 μm. If the material was, however, heavily doped (with 2 at% La), the grain diameter did not exceed 1 μm.

Drofenik (1987) examined how the partial pressure of oxygen affected the grain growth of doped barium titanate. He sintered $BaTiO_3$ specimens doped with Sb_2O_3 and Nb_2O_5 (from 0.20 to 1.0 at%) and added with 0.6 wt% SiO_2 and 4 mol% TiO_2 at a temperature of 1610 K at various partial pressures of oxygen (from 0.4 MPa to 20 Pa; at small partial pressures, the oxygen was mixed with nitrogen). At low oxygen pressures ($\ll 0.02$ MPa), neither grain growth nor dopant enrichment were inhibited; it was possible to increase the dopant concentration to above 0.3%.

The diameter of the $BaTiO_3$ grains also depends on the excess of Ti above the stoichiometric proportion. Ihrig (1978) has shown that, in a material doped with 0.2 mol% Sb (which is a donor-type additive), the grain diameter increases from 10 μm, at a Ti/Ba ratio equal to 1.0, to 40 μm, at Ti/Ba = 1.02, and then decreases to about 20 μm, at Ti/Ba = 1.05. The addition of Sb also increases the amount of the intergranular phase at the $BaTiO_3$ grain boundaries. A small amount of eutectic phase ($Ba_6Ti_{17}O_4$–$BaTiO_3$) was reported to cause discontinuous grain growth above a temperature of 1585 K (Hennings et al., 1987).

Lin et al. (1990) report that, at a sintering temperature of 1620 K, they have obtained large grains (30–50 μm) of La-doped $BaTiO_3$ with an excess of Ba (Ti/Ba= 0.99), but small grains (1–3 μm in samples with an excess of Ti (Ti/Ba = 1.01). Howng and McCutcheon (1983) examined how an addition of Ca affects the grain diameter in a Ba–Ti–Pb–Ca ceramic doped with Y, and found that an addition of 7 mol% Ca reduces the average grain diameter by about 35%, i.e., to about 15 μm.

The grain diameter of PTCR materials may be adjusted by controlling the formation of the liquid phase during the sintering process. When producing PTCR components intended for stable operation at a relatively high voltage, such as heating elements operating at a constant temperature, it is important that their microstructure is uniformly fine-grained, which ensures that they are not liable to electric breakdown.

Typical components of the liquid phase include TiO_2, SiO_2 and Al_2O_3, which form an eutectic with $BaTiO_3$. TiO_2 forms an eutectic with $BaTiO_3$ at a temperature of 1595 K (Rase and Roy, 1955a). The eutectic temperature may be reduced to 1533 K by adding a small amount of SiO_2 (Rase and Roy,

1955b) and further to 1513 K by adding a certain amount of Al_2O_3 (Matsuo and Sasaki, 1971). Al-Allak et al. (1988) have shown that an addition of 0.55 mol% Al_2O_3 increases the size of the $BaTiO_3$ grains by about 2 µm, i.e., to about 8 µm.

Schmelz and Meyer (1982) observed an exaggerated growth of the $BaTiO_3$ grains. They suggest that this effect may be due to certain reactions, such as surface diffusion, that occur in the solid phase and, thus, it does not depend on the presence of the liquid phase. The growth nuclei may be here, for example, twinned $BaTiO_3$ grains. This hypothesis has been confirmed by Eibl et al. (1987).

As in the case of ZnO varistors, investigators differed initially in their opinions about whether a separate intergranular phase did occur in PTCR materials and if so, whether it contributed to the occurrence of the PTCR effect. However, TEM (transmission electron microscope) examinations soon proved that, in a fully sintered PTCR ceramic material of good quality, no intergranular layer wider than 2–10 nm occurred (Haanstra and Ihrig, 1980). This phase only forms temporarily when the temperature is high during the sintering process and, after cooling, it only remains where three or more $BaTiO_3$ grains are in contact. Since the results obtained by these investigators have definitively discarded the hypothesis that the PTCR effect is due to the presence of an intergranular phase, the mechanisms based on this hypothesis will not be discussed in this book.

Jonker (1981) found that, above 1590 K, a liquid intergranular phase composed of $BaO \cdot 3TiO_2$ is formed, and that the presence of this phase enhances the sintering process. The intergranular phase forms from the excess (with respect to the stoichiometric proportion) TiO_2, and after the material has been cooled down, it remains in a vitreous state at triple-grain junctions. If $BaTiO_3$ has been doped, the intergranular phase contains dissolved oxide dopants. When doping the material with metal oxides, with the intention that they should introduce additional acceptor levels on the surfaces of the $BaTiO_3$ grains, the presence of the intergranular phase facilitates the distribution of the oxides over the grain surfaces.

Haanstra and Ihrig (1980) report on their TEM examinations of ferroelectric $BaTiO_3$ grains and the boundaries between them. They found that both the grains and the boundaries show a domain structure, and that the individual domains touch each other at an angle of 90°.

Young-Sung Yoo et al. (1987) found that the rate at which the material is heated to the sintering temperature affects its properties, especially when the material is doped with an excess of TiO_2. At lower heating rates (0.5 K/min), the porosity and the average grain size were greater in the central portion of the sample than in its outer layer; in macroscopic terms the sample

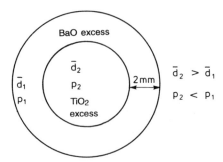

Fig. 4.6. The macroscopic heterogeneity of the porosity and average grain diameter of a doped $BaTiO_3$ sample slowly heated to the sintering temperature; d—the average grain diameter, p—the porosity (after Young-Sung Yoo et al., 1987)

was heterogeneous (Fig. 4.6). At higher heating rates (5 and 50 K/min), the whole sample had a homogeneous microstructure. The investigators explained the observed heterogeneities in terms of the nonuniform distribution of the TiO_2 excess, which appeared in a liquid phase during the sintering operation. When the material was heated at a slower rate, the liquid phase tended to accumulate in the centre of the sample, where it stimulated the growth of grains. As a result, the outer layer of the sample was enriched in BaO, whose effect was reverse to that of TiO_2. These effects occurred because of the absence of a liquid phase during the initial sintering stages, and vanished after the heating rate was markedly increased. Microscopic examination of the samples that had been heated at a slower rate showed that in their central portions the concentration of Ti dissolved in the $BaTiO_3$ grains was increased, and that no intergranular phase was present at the grain boundaries. Similar results have been reported earlier by Mostaghaci and Brook (1983).

To assess how the properties of the PTCR materials depend upon the parameters of their fabrication process, in particular the parameters of the sintering operation, we need to know the mechanisms that then operate. These will be discussed in Section 4.4, where the reader will also find further indications as to the parameters of the fabrication process, the criteria used for choosing an appropriate chemical composition of the ceramic material, and the effects of the individual constituents upon its final properties.

Kuwabara (1980, 1981a, b, c, 1983b) was the first to begin studies on new PTCR materials of greater porosity, the resistivity of which varied by seven orders of magnitude. These materials did not contain dopants such as Mn and Cr, that would introduce additional acceptor levels (see Section 4.4). Two materials of the same composition $Ba_{0.998}Sb_{0.002}TiO_3$ were prepared either from powder produced by the thermal decomposition of $BaTiO(C_2O_4) \cdot 4H_2O$ (barium titanyl oxalate) or by an appropriate solid state reaction between $BaCO_3$ and TiO_2. The fabrication process was typical of ceramic technology, except that special attention was paid to achieving a density below 95% of the theoretical density. In the first experiment, water was avoided during the entire process. After being mixed in ethyl alcohol and subjected to calcination at a temperature between 1170 and 1470 K, the materials were pressed at various pressures amounting to 200 MPa, to form pellets (10 mm in diameter and 2.5 mm thick), and sintered in air at a temperature of 1620 K for 1 to 2 hours. The temperature was raised at a rate of 10 K/min, and then decreased at a rate of 100 K/min or more. The densities of the samples thus obtained ranged from 60 to 90% of the theoretical density and the grain diameters ranged from 2 to 5 μm. In–Ga electrodes were deposited on the faces of the samples. A high temperature-resistivity coefficient was only obtained in samples with small $BaTiO_3$ grains. Figure 4.7 shows how the PTCR effect depends on the relative density (calculated relative to the theoretical density value which is 6.01 g/cm^3) of the samples.

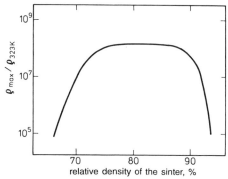

Fig. 4.7. The variation of the PTCR effect as a function of the relative density of porous sintered materials (after Kuwabara, 1983b; reprinted by permission of the American Ceramic Society)

Kuwabara (1983e) also described the process of fabrication of a PTCR resistor pressed out of a $BaTiO_3$ powder that contained Sb_2O_3 and 10–20% of metallic indium. The $BaTiO_3$ powder was produced from barium titanyl oxalate with an addition of Sb_2O_3, which after mixing was calcined in air at a temperature of 1170 K for 1 hour. Then, to increase its conductivity, the powder was heated at a temperature of 1570 K for 1 hour. The powder thus obtained (with a grain diameter about 50 μm) was mixed with indium at 470 K (the melting temperature of indium is 429 K), and appropriate pellets (10 mm in diameter, 2.5 mm thick) were pressed out of it under a pressure of about 140 MPa. The electrodes were made of a silver paste, fired at about 620 K for 10 min. The resistivity of these resistors varied with temperature by as much as four orders of magnitude. The same ceramic material may be used for fabricating, by the thick-film technique, constant-temperature heating devices to be deposited on curved surfaces.

Ki Hyun Yoon and Eun Hong Lee (1987) proposed fabricating PTCR resistors by synthesis of the material in melted KCl. They mixed $BaCO_3$, TiO_2 and Sb_2O_3 with KCl in ethyl alcohol and then subjected the mixture to calcination at a temperature of 1070 K. The weight proportion of the KCl to the other constituents was varied from 0 to 1. The calcined material was crushed and KCl was washed out with deionized water until $AgNO_3$ ceased to react to Cl ions. The powder thus obtained was then pressed, without adding a plasticizer, under a pressure of 50 MPa, and sintered at a temperature of 1620 K for 2 hours. The resistivity of the final material appeared to increase with increasing temperature by three orders of magnitude.

4.3 ELECTRICAL PROPERTIES OF PTCR MATERIALS

The current-voltage characteristic of a PTCR material may be considered to be a combination of the characteristics of two components, connected in parallel, namely, an ideal PTCR resistor (which has a linear characteristic) and a varistor. How these two characteristics can be combined is shown in Fig. 4.8, which, however, does not take into account the selfheating effect that occurs in the PTCR material due to the flow of an electric current through it. As the temperature increases, the resistance of the material sharply increases and, thus, the current decreases. If the admissible voltage is not exceeded, the current and temperature can reach stable values. The actual current-voltage characteristic of a thermistor is shown in Fig. 4.9.

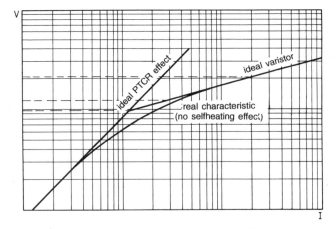

Fig. 4.8. The current-voltage characteristic of a real PTCR resistor shown as a superposition of the characteristic of an ideal PTCR resistor and that of an ideal varistor (redrawn from the Philips Catalogue, 1984)

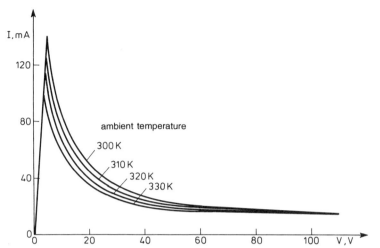

Fig. 4.9. The current-voltage characteristic of a real PTCR resistor at various ambient temperatures (redrawn from the Philips Catalogue, 1984)

Up to certain value of the voltage this characteristic follows Ohm's law. With a further voltage increase, however, the temperature of the thermistor increases, thereby increasing its resistance. As a result, the current decreases to a constant value. The shape of the characteristic is, of course, dependent on the ambient temperature.

The porous PTCR materials obtained by Kuwabara using the technique described in the preceding section, when examined in air in the temperature range from 370 to 520 K, gave a resistivity change of more than seven orders of magnitude, when the temperature increased at a rate of 3 K/min. A typical curve is shown in Fig. 4.10. This figure also shows the temperature variation of the dielectric constant, calculated from the capacitance measurements at a frequency of 1 kHz. In view of the high thermal coefficient of resistance, these materials may be expected to find large-scale application in the future.

That potential barriers are present at the boundaries of $BaTiO_3$ grains has been confirmed experimentally by Gerthsen and Hardtl (1963), who used decoration techniques, Rehme (1968), who performed emission electron microscope examinations, Haanstra and Ihrig (1977), who examined the voltage contrast on a scanning electron microscope, and Ihrig and Klerk (1979), who employed cathodoluminescence at room temperature.

Gerthsen and Hoffmann (1973) measured the variation of the capacitance of a single $BaTiO_3$ grain boundary as a function of a direct voltage applied to it. The samples had the form of thin (0.01 mm) plates cut from a sintered ceramic. Ni electrodes were deposited on the surfaces of the neighbour-

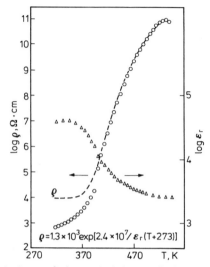

Fig. 4.10. Typical variations of the resistivity and dielectric constant in porous PTCR resistors (after Kuwabara, 1983a; reprinted by permission of the American Ceramic Society)

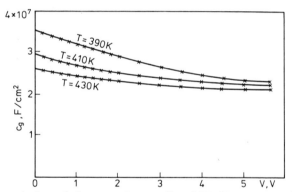

Fig. 4.11. The variation of the capacitance C_g of PTCR resistors (1 cm^2 of the grain boundary) as a function of the dc voltage applied at various temperatures above the Curie point (after Gerthsen and Hoffmann, 1973; reprinted with kind permission from Pergamon Press Ltd, Headington Hill Hall, Oxford OX3 0BW, UK)

ing grains. Figure 4.11 shows the variation of the capacitance, as measured by these investigators. Based on these results and on the results obtained from the measurements of the current–voltage characteristics of single grain boundaries, Gerthsen and Hoffmann concluded that an intergranular phase occurs at the grain boundaries and that the PTCR effect is due to this phase. In view of the fact, however, that later examinations of best performance PTCR materials did not confirm the presence of this phase, we may explain the Gerthsen and Hoffmann results by assuming that the PTCR effect observed by them was due to an acceptor film adsorbed on the grain surfaces.

Nemoto and Oda (1980, 1981) also examined the properties of single grain boundaries in a PTCR material, using a micromanipulator whose electric probes were made of a 25 µm-diameter wire. They measured the changes of resistivity with temperature and the current-voltage characteristics of the boundaries. They concluded that the origins of the PTCR effect lay at the $BaTiO_3$ grains boundaries. They also confirmed the observation that below the Curie temperature the resistivity of the grain boundaries was greater than that inferred from the resistivity of the grain bodies. Contrary to other investigators, they considered the Heywang model (discussed in Section 4.4) to be incorrect, and they suggested that the PTCR effect was a result of the decreased mobility of the electrons captured at the $BaTiO_3$ grains rather than caused by changes of the heights of the potential barriers.

Another useful technique, which distinguishes between the bulk and grain boundary electrical properties is ac impedance spectroscopy (Macdonald, 1987). Tseng and Wang (1990) give impedance spectra for high-Curie-point barium lead titanate at various ambient temperatures.

The range of application of PTCR materials is determined by their Curie temperatures. In $BaTiO_3$, this temperature may be adjusted by appropriate doping. Kuwabara and Kumamoto (1983f), Kuwabara et al. (1985) and S.-H. Wang et al. (1990) adjusted the Curie temperature of this material by introducing Pb and Sr ions into the sites of the Ba crystal lattice. Kuwabara and Kumamoto obtained materials whose Curie temperature, as can be seen in Fig. 4.12, ranged from 330 to 630 K. The materials whose Curie temperature was the lowest (marked with the small arrows in the figure) showed the greatest resistivity at room temperature. If Pb replaces Ba in the lattice sites, the Curie temperature increases, whereas Sr in place of Ba, and Zr or Sn replacing Ti in the lattice sites decrease Curie temperature. In the material with the Curie temperature equal to 630 K, the changes of resistivity exceeded four orders of magnitude.

Kuwabara (1990) has recently attempted to fabricate PTCR thermistors of $PbTiO_3$ doped with Nb, Bi and BN. This material exhibits a high Curie point (750–760 K) with an increase of resistivity by as much as 3 orders of magnitude; it was however found to be strongly unstable. Chin and Lou (1991) have also prepared a porous single phase $PbTiO_3$ PTCR material which shows a two orders of magnitude increase in resistivity. They did not however examine its stability.

Kuwabara (1981b, 1983b) examined how the magnitude of the PTCR effect (measured as the ratio of the maximum value of the resistivity to its value at a temperature of 320 K) varies as a function of the excess addition

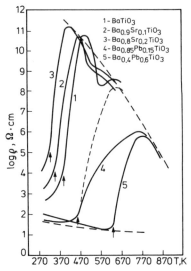

Fig. 4.12. Effect of the chemical composition of the PTCR material upon its Curie temperature (denoted by the arrows); (after Kuwabara and Kumamoto, 1983f; reprinted by permission of the American Ceramic Society)

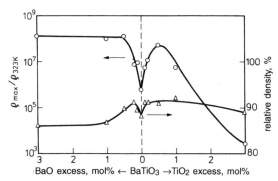

Fig. 4.13. The intensity of the PTCR effect and the relative density of the sintered material as functions of excess of BaO and TiO_2 (after Kuwabara, 1981b; reprinted by permission of the American Ceramic Society)

of BaO and TiO_2 in a porous material designed by him. This variation together with the variation of the relative density (calculated with respect to the theoretical density value of 6.01 g/cm^3) of this material is shown in Fig. 4.13. An addition of more that 3 mol% TiO_2 resulted in the formation of a substantial amount of the liquid phase and stimulated the growth of grains. According to Kuwabara, this inhibited the PTCR effect.

Ihrig (1977, 1978) determined the variation of the resistivity of an Sb-doped material as a function of the excess of TiO_2 (Fig. 4.14). He measured the variations of both the maximum (at a temperature from 470 to 520 K) and the minimum (at a temperature of 320 K) resistivity value. It is interesting to note that these materials preserved the ability to increase sharply their resistivity even though the excess of TiO$_2$ was 50%, and irrespective of the presence of the intergranular phase. Contradicting the results obtained by Ihrig (1978), Basu and Maiti (1987) have shown (by measuring impedance) that, when present at the BaTiO$_3$ grain boundaries, the intergranular phase increases the room-temperature resistivity of the material.

Kuwabara (1981b) examined the ageing effects that occurred in the porous PTCR material designed by him. He stored the samples in dry (relative

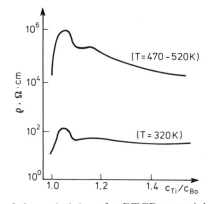

Fig. 4.14. The variation of the resistivity of a PTCR material as a function of the excess of TiO$_2$ (after Ihrig, 1978; reprinted by permission of the Ceramic Forum International, Berichte der Deutschen Keramischen Gesellschaft)

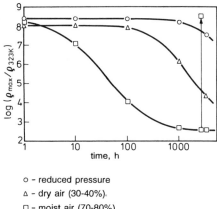

○ – reduced pressure
△ – dry air (30-40%).
□ – moist air (70-80%)

Fig. 4.15. Typical ageing effects occurring in porous PTCR resistors operated in various environments. After firing the resistors in oxygen at 1370 K, their parameters recover to the original values (after Kuwabara, 1981b; reprinted by permission of the American Ceramic Society)

humidity of 30–40%) and wet (relative humidity of 70–80%) air under atmospheric pressure and reduced pressure (1.3 Pa). How the ageing process affects the PTCR effect is shown in Fig. 4.15. When the ageing was carried out in wet air under atmospheric pressure, the PTCR effect was drastically reduced, due chiefly to an increase of the low-temperature resistivity $\varrho_{320\ \text{K}}$. It should be noted that the air could easily penetrate Kuwabara's material because of its high porosity.

When heated below the Curie temperature, samples made of a porous PTCR material ($Ba_{0.398}Pb_{0.600}Sb_{0.002}TiO_3$) of high Curie temperature (T_C = 630 K) exhibited a resistivity hysteresis (Kuwabara et al., 1988). This can be seen in Fig. 4.16. Kuwabara et al. related this effect to the observed increase in resistivity described by

$$\log \varrho = \log \varrho_0 + B\sqrt{t}, \qquad (4.2)$$

where t is time, ϱ_0 the initial resistivity at $t = 0$, and B a constant. The variations of values of $\log \varrho_0$ and B as functions of temperature are shown in Fig. 4.17. Kuwabara et al. then prepared another ceramic material of the same composition and Curie temperature, but not exhibiting the thermal hysteresis. This was achieved thanks to its homogeneous fine-grained microstructure.

Kuwabara and Inoue (1983c) examined the effect of various atmospheres upon the conduction of porous PTCR ceramics. They measured the variation of resistivity of the material at room temperature in the following atmospheres: air, O_2, N_2, CO, CO_2, CH_4, $N_2 + H_2O$, $N_2 + C_2H_5OH$, and $N_2 +$ isopropyl alcohol. The last three gas mixtures were produced by evapo-

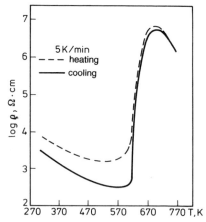

Fig. 4.16. Hysteresis of the resistivity of a PTCR resistor observed below the Curie temperature (after Kuwabara et al., 1988; reprinted by permission of the American Ceramic Society)

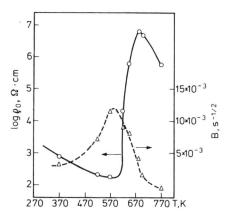

Fig. 4.17. Variations of the term $\log \varrho_0$ and the factor B in equation (4.2) as functions of temperature (after Kuwabara et al., 1988; reprinted by permission of the American Ceramic Society)

rating from the liquid state and transporting the vapour using the N_2 carrier. Kuwabara and Inoue also measured the variation of the resistivity with temperature. They found that, above the Curie point, reducing environments drastically inhibited the PTCR effect. Samples degraded in a reducing atmosphere could recover to their initial properties by exposing them to oxygen at a temperature higher than 570 K. When the material was placed in a CO atmosphere at 620 K, its resistivity decreased by three orders of magnitude during 15 s, which makes it a good choice for constructing a gas sensor sensitive to CO or to alcohols.

4.4 OPERATING MECHANISM OF PTCR MATERIALS

In contrast to the differing views of investigators about the operating mechanism of ZnO-based materials, their opinions about the operating mechanism of PTCR materials are almost in agreement. The model proposed by Heywang in 1961 is still considered to be basically correct, although it has been substantially improved since that time.

4.4.1 Energy band structure of the grain boundaries above the Curie temperature

Soon after the PTCR materials were discovered, Heywang (1961, 1971) constructed a model describing their behaviour. Even now, the assumptions underlying the Heywang model are considered to be true. The model represents the energy band structure of the boundary of a $BaTiO_3$ grain, which is responsible for the rapid increase of electrical resistance that is observed at certain temperatures (Fig. 4.18). At the grain boundary, surface acceptor

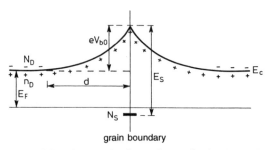

Fig. 4.18. Potential barrier at the boundary of a barium titanate grain

states are formed due, for example, to oxygen adsorbed at the interface. Above the Curie temperature, the presence of these states results in potential barriers forming at the boundary. As we have shown in Subsection 2.2.2, the height V_b of such a barrier is inversely proportional to the value of the dielectric constant ε of BaTiO$_3$, and is given by

$$V_b = \frac{eN_S^2}{2\varepsilon\varepsilon_0 N_D} \qquad (4.3)$$

where N_S is the concentration of the surface states, and N_D the concentration of the donor levels.

The presence of a potential barrier of height $E_b = eV_b$ results in the resistivity of the material increasing by the factor $\exp(E_b/kT)$. The height of the barrier, as reported by Heywang (1964) who calculated it based on measurements of the electrical conductivity of the material at the Curie temperature, was $E_b = 0.5$ eV. It should be noted that Heywang only took into account thermally activated carrier transfer over the barrier. The behaviour of PTCR resistors has been explained in terms of the ferroelectric–paraelectric transition that occurs at the Curie temperature. According to the Curie–Weiss relationship (equation (4.1)), the dielectric constant decreases with increasing temperature above the Curie point (Fig. 4.3) and, thus, according to equation (4.3), the height of the potential barrier increases. Because of the exponential variation of the electrical conductivity with the barrier height, we observe the sharp increase of the resistivity. Jonker (1964) in addition takes into account the fact that the concentration n_S of the electrons trapped at the interface varies with temperature according to the equation

$$n_S = \frac{N_S}{1 + \exp[(E_F + eV_{b0} - E_S)/kT]} \qquad (4.4)$$

where $E_F = kT\ln(N_D/n_D)$ is the Fermi level, E_S the position of the trap levels, and eV_{b0} the height of the potential barrier, which, at the Curie temperature, is equal to zero. Using this equation and assuming that the electron mobility is low and the electric currents are small, we can calculate the resistance of the barrier from

$$R = \frac{d}{n_D e \mu} \frac{kT}{eV_{b0}} \exp\left(\frac{eV_{b0}}{kT}\right) \qquad (4.5)$$

where d is the width of the barrier; the surface area of the barrier has been taken to be equal to 1 cm^2.

From this equation we can see that the resistivity depends strongly upon temperature. Jonker has calculated that, at the Curie point when the barrier height is equal to zero, the surface (trap) states are positioned far below the Fermi level and are almost entirely ionized ($n_S = N_S$, cf. equation (4.4)). As the temperature increases, the trap levels move to higher positions until they reach the Fermi level. At this moment, the height of the potential barrier is at its maximum: $eV_b = E_S - E_F$. The above reasoning explains why the resistivity–temperature variations observed in various experiments differ from one another (Fig. 4.19).

In order to obtain a possibly high potential barrier, we must have a high concentration of surface states. It follows from equations (4.3) and (4.5) that at a high concentrations of the surface states N_S, the resistivity reaches a maximum above the Curie point and then it slowly decreases. At low N_S, the resistivity continuously increases with temperature, within the temperature range of several hundred Celsius degrees.

Summing up we may say that the rapid increase of the resistivity of PTCR materials that occurs above the Curie temperature is due to the variation of the dielectric constant and the concentration of the surface states, and also due to the ferroelectric-paraelectric transition that takes place when the temperature passes through the Curie point.

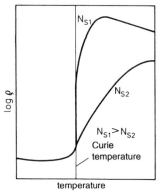

Fig. 4.19. Two types of resistivity-temperature curve corresponding to two different concentrations of surface states N_S (after Jonker, 1964; reprinted with kind permission from Pergamon Press Ltd, Headington Hill Hall, Oxford OX3 0BW, UK)

4.4.2 Energy band structure of the grain boundaries below the Curie temperature

What remain to be explained are the rapid increase of the resistivity at the moment when the temperature passes through the Curie point, and the fact that the resistivity does not increase when, below the Curie point, the dielectric constant again decreases. Jonker (1964) assumed that theses two facts are associated with the passing of the crystal into a ferroelectric state, as a result of which almost all the surface states are compensated and, thus, the potential barrier does not form. After the crystal passes into the ferroelectric state, each of the grains of semiconducting $BaTiO_3$ acquires a domain structure. The free energy of such a material will be at a minimum when the domain structures of neighbouring $BaTiO_3$ grains are ideally matched to one another. Since, however, we have two grains of different crystallographic orientation, such a match between their structures cannot occur (Fig. 4.20). A certain degree of matching can only be achieved by inducing stresses on the surfaces of the grain boundaries.

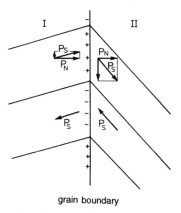

Fig. 4.20. Diagram illustrating the matching between the ferroelectric domains at the boundary of a barium titanate grain and the formation of space charge layers; P denotes polarization (after Jonker, 1964; reprinted with kind permission from Pergamon Press Ltd, Headington Hill Hall, Oxford OX3 0BW, UK)

In semiconductor crystals, the domains may be matched due to the compensation of the surface states by electric charges; a structure composed of charged layers forms then as shown in Fig. 4.20. These layers rapidly form upon passing down through the Curie temperature (during the cooling cycle), resulting in the existing potential barriers being suppressed immediately: below the Curie point, no effect of the barriers is observed. On about 50% of the total surface of the grain boundaries a negative charge occurs, compensating the ionized donor levels. The absence of potential barriers on these boundaries permits conducting paths to develop through the ceramic, thanks to which its resistivity at room temperature is low.

The temperature range near the Curie point is the operating range of PTCR resistors. Ideally, we wish to have a material that below the Curie temperature behaves as a metal, and above this temperature as an insulator. At $T < T_C$, the conductivity of this material should only be limited by the number and mobility of the carriers. Ihrig (1983), however, suggests that potential barriers also exist at these temperatures. His hypothesis has been confirmed by observations of cathodoluminescence: the luminescent intensity decreases at the grain boundaries (Hoffmann, 1973; Brauer, 1974). Ihrig further suggests that the structural stresses induced as a result of the matching between the domain structures of two neighbouring grains are chiefly relaxed by mechanical deformation, and, thus, not only by the movements of the surface charges. Thanks to the presence of the potential barriers that remain at certain grain boundaries, the resistivity of the ceramic material below the Curie temperature is several times greater than the value that can be expected judging from the resistivity of the interior of the $BaTiO_3$ grains.

According to Kulwicki and Purdes (1970), the height of the potential barrier below the Curie temperature is given by

$$eV_b = \frac{e^2(N_S - 2P_n)^2}{8\varepsilon\varepsilon_0 N_D} \quad (4.6)$$

where P_n is a polarization component, perpendicular to the surface of the grain boundary. We can see that the surface states may be compensated as a result of polarization.

A knowledge of the mechanism of the PTCR effect permits designing the manufacturing processes so as to control the properties of the PTCR materials by adjusting the properties of the elements of their structure such as

—the interior of the $BaTiO_3$ grains,

—the surface of the samples (in order to control the formation of surface potential barriers),

—the boundaries of the $BaTiO_3$ grains.

4.4.3 Electrical conductivity of the interior of barium titanate grains

4.4.3.1 Defect structure of $BaTiO_3$ doped with La

As mentioned earlier, the resistivity of the interior of barium titanate grains can be controlled by appropriate doping and/or by adjusting the composition of the atmosphere in which the sintering process is carried out. An example of the dopant may be La, which is substituted for Ba in the crystal lattice. La^{3+} ions are of almost the same size as the Ba ions and, thus, they do not tend to substitute for the Ti ions, which are much smaller. The composition thus obtained is

$$(Ba_{1-x}La_x)(Ti^{+4}_{1-x}Ti^{+3}_x)O_3 \qquad (4.7)$$

When doping in this way, we do not increase the number of oxygen vacancies in the BaTiO$_3$. The excess charge introduced by doping is compensated either by the electrons passed to the conduction band, or by the formation of a Ba vacancy. Daniels et al. (1978/79) examined how the conductivity of BaTiO$_3$ doped with La varies with the oxygen partial pressure at a temperature of 1470 K, under the conditions of thermodynamic equilibrium of point defects. They have found that, at very high oxygen pressures, undoped BaTiO$_3$ shows p-type conductivity, which changes into n-type as the oxygen pressure decreases. When the material was doped, its conductivity was always n-type. They observed three different regions in the $\sigma(P_{O_2})$ curve, where σ denotes the conductivity, corresponding to the three different pressure regions that occurred during the experiment.

Daniels et al. have constructed a model describing the formation of defect equilibria, based on the assumption that the only defects are oxygen vacancies (V$_O$, V$_O^\bullet$, V$_O^{\bullet\bullet}$) and barium vacancies (V$_{Ba}$, V$'_{Ba}$, V$''_{Ba}$). Ti vacancies,

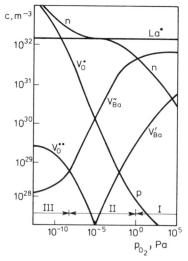

Fig. 4.21. Concentration of point defects in BaTiO$_3$ doped with lanthanum at 1470 K and varying oxygen partial pressure. The predominant defects (i.e., those which contribute most to the material conduction) are barium vacancies in region I, La ions in region II and oxygen vacancies in region III (after Daniels et al., 1978/79)

interstitial ions and effects that can occur between defects of great concentration have been neglected in the model. The electroneutrality condition then takes the form

$$n + [V'_{Ba}] + 2[V''_{Ba}] = p + [V^{\bullet}_O] + 2[V^{\bullet\bullet}_O] + [La^{\bullet}] \qquad (4.8)$$

Daniels et al. have also described the partial reactions of the charge and mass conservation and, based on experimental data, calculated appropriate equilibrium constants. They have examined how the conductivity of $BaTiO_3$ doped with La varies with the oxygen partial pressure, and the relation calculated by them is in good agreement with experiment. They also give the equilibrium curves of defect concentration (Fig. 4.21). They have found that, when the conductivity of undoped $BaTiO_3$ is chiefly a function of the concentration of oxygen vacancies, the three regions of the $\sigma(p_{O_2})$ curve (Fig. 4.21) are determined by the electroneutrality conditions

I. $\quad 2[V''_{Ba}] \approx [La^{\bullet}] \quad (p_{O_2} = 10^0 \text{--} 10^5 \text{ Pa})$,

II. $\quad n \approx [La^{\bullet}] \quad (p_{O_2} = 10^{-7} \text{--} 10^0 \text{ Pa})$,

III. $\quad n \approx [V^{\bullet}_O] \quad (p_{O_2} = 10^{-12} \text{--} 10^{-7} \text{ Pa})$.

In region III, the variation of the conductivity as a function of the oxygen partial pressure is analogous to its variation in undoped $BaTiO_3$, whereas, in region I, La^{+3} ions are compensated by barium vacancies and, in region II, by electrons. Figure 4.22 shows, according to Daniels et al., the positions of the energy levels corresponding to various defects present in the $BaTiO_3$.

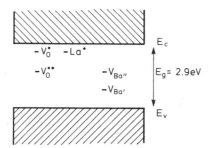

Fig. 4.22. Positions of the energy levels of point defects in the forbidden band of $BaTiO_3$ (after Daniels et al., 1978/79)

The properties of PTCR materials are chiefly determined by the conditions that prevail during the process of their sintering. The rate of changing the temperature during this process and the atmosphere in which it is carried out should be selected according to the indications that follow from the kinetics of the diffusion of point defects.

4.4.3.2 The kinetics of defect diffusion

Daniels et al. (1978/79) investigated the kinetics of the diffusion of defects in BaTiO$_3$ by measuring the changes in its conductivity after an abrupt change of the oxygen partial pressure. Using specimens of various thicknesses and grain sizes, they determined which diffusion mechanisms predominate and whether the changes in conductivity are due to diffusion or to reactions that occur on the grain surfaces. In the undoped material, within pressure region I (Fig. 4.21—the highest pressures), it is the barium vacancies that predominate and their diffusion determines the conductivity values. The diffusion coefficients of these vacancies are much lower than those of oxygen vacancies. Based on these measurements, Daniels et al. describe the mechanism of the formation of barium vacancies. Their description greatly contributes to the explanation of the PTCR mechanism in atomic terms.

In the manufacture of PTCR materials, it is common practice to add an excess (with respect to the stoichiometric composition) of TiO$_2$ in order to enhance the sintering process. Daniels et al. suggest that the phase BaTi$_3$O$_7$ is then formed at the grain boundaries. At high temperatures, the Ba ions may diffuse from the crystal lattice into this phase, so that it may be difficult to identify this phase after sintering. The diffusing Ba ions leave barium vacancies behind them in the titanium lattice according to the equation

$$\text{BaTi}_3\text{O}_7 + 2\text{Ba(lattice)} + 2\text{O(lattice)} \rightleftarrows 3\text{BaTiO}_3 + 2V_{\text{Ba}} + 2V_{\text{O}} \quad (4.9)$$

Depending on the rate of cooling from the high temperature, we obtain various structures of the BaTiO$_3$ grains. Figure 4.23 shows the variation of the conductivity of BaTiO$_3$ specimens, doped with La, which were cooled from various temperatures at different oxygen partial pressures. At high temperatures (above what is known as the inversion temperature), the total concentration of [La^{tot}] is higher than the concentration [V$_{\text{Ba}}^{\text{tot}}$]. If the speci-

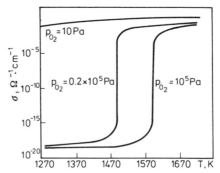

Fig. 4.23. Variation of the conductivity of a PTCR material cooled from the sintering temperature at different oxygen partial pressures (fast cooling); redrawn from Daniels et al., 1978/79

men is cooled rapidly, lanthanum is not fully compensated and the material exhibits a high conductivity. In practical manufacturing processes, the cooling rates are lower and vary from place to place within the sample because of the nonuniform temperature distribution. In effect, the state of the material departs from thermodynamic equilibrium and is determined by the kinetics of the diffusion of point defects whose movements do not keep pace with the decreasing temperature. Thus, in practice, we have a diffusion 'front' of barium vacancies formed at the grain boundaries, and this 'front' moves towards the bulk of the grain (Fig. 4.24). In near-boundary regions, all the La ions are compensated and layers of high resistivity form there.

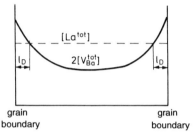

Fig. 4.24. A schematic representation of the defect concentrations in La-doped $BaTiO_3$, due to which the PTCR effect occurs. When the total concentration $[La^{tot}]$ does not exceed $2[V_{Ba}^{tot}]$, all the conduction electrons introduced by doping are trapped; l_D is the width of the space charge layer (after Daniels et al., 1978/79)

It has been found that the barium vacancies are acceptor states which create potential barriers at the $BaTiO_3$ grain boundaries, and are responsible for the positive sign of the temperature coefficient of resistivity (the PTCR). According to the PTCR models discussed above, in manufacturing PTCR resistors we should tend to produce a thin layer of high resistivity, whereas the bulk of the grain should maintain a high conductivity. Pure, undoped $BaTiO_3$ does not show the PTCR effect, since the conditions prevailing there do not promote the formation of barium vacancies and, thus, the acceptor states. The defects that predominate in such material are oxygen vacancies.

4.4.3.3 $BaTiO_3$ doped with Sb ions

As mentioned earlier, in doping the Ba crystal lattice with ions of higher valencies other ions, apart from La ions, are also used. Another frequent additive is Sb_2O_3. In experiments (Eberspachel, 1962; Schmelz, 1969; Heywang, 1970) on the behaviour of $BaTiO_3$ doped with antimony oxide, two cases were examined: $Ba(Ti_{1-x}Sb_x)O_3$ (a deficiency of Ti) and $(Ba_{1-x}Sb_x)TiO_3$ (a Ba deficiency). With the first material, the investigators did not observe any increase in conductivity, and the results of examinations of the lattice constants and neutron diffraction suggested that pairs of Sb^{+3} and Sb^{+5} ions are formed there. The other material exhibited an increase of conductivity, but a single-phase material was only obtained at small Sb contents. At higher Sb concentrations, the Sb ions substituted on both Ba and Ti sites, but Ti^{+3} ions did not form:

$$(Ba^{+2}_{1-u}Sb^{+3}_{u})(Sb^{+3}_{u+v}Sb^{+5}_{v}Ti^{+4}_{1-u-2v})O_3 \qquad (4.10)$$

The resulting excess titanium ions occurred in the vitreous intergranular phase of the sintered material.

4.4.3.4 Critical doping level

It would seem that in order to increase the conductivity of $BaTiO_3$ grains and, at the same time, reduce the thickness of the compensated layer as much as possible, the concentration of donors in the Ba lattice should be increased up to the solubility limit of a given dopant. In practice, however, the conductivity of the material decreases after a relatively low threshold donor content of about 0.3 mol% is exceeded. This effect, known as the doping anomaly, is illustrated in Fig. 4.25.

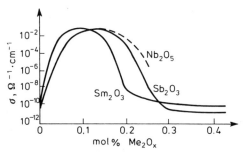

Fig. 4.25. The decrease of the conductivity of $BaTiO_3$ ceramic after the critical doping level is exceeded (after Heywang, 1971; reprinted by permission of Chapman & Hall)

When, however, the dopant concentration exceeds a certain characteristic value, the average size of the ceramic grains decreases. In view of the fact that the width of the potential barrier (the depletion layer) is of the order of 1 μm, we may expect that when the grains of the basic phase are about this size, there will be no free carriers within them.

What remains to be explained is the mechanism which inhibits the growth of the grains. Practically, the critical doping concentration is independent of the kind of additive and it does not even depend on whether the additive atoms are placed in the Ba sites (as is the case with rare earth metals, such as Sb, Y and Bi) or in Ti sites (the case with Nb and Ta), cf. Hanke and

Schmelz (1982) and Hanke (1973). When, in addition to donors, some acceptors are introduced into the material (e.g., certain impurities introduced during the manufacturing process), the critical doping level shifts towards higher concentrations. These acceptor additives may include Fe, Co, Cr and Mn. After adding 1.2 mol% Co, for example, the critical concentration of Nd shifts to between 0.3 and 1.5 mol%. The observed changes of the critical concentration may also be attributed to the acceptor impurities contained in the raw materials. As for the donor additives, their critical concentration level depends on the sintering temperature (it increases as this temperature increases), the kind of the atmosphere in which the sintering process is carried out, and on the oxygen partial pressure (the critical concentration level increases as the oxygen partial pressure decreases).

Hanke and Schmelz (1982) have proposed a model that explains these anomalies. It follows from the defect equilibria, described earlier, that in sintered $BaTiO_3$, some of the carriers are compensated by the acceptor levels that result from barium vacancies. In the reducing atmosphere in which the sintering is carried out, donor oxygen vacancies also form. By considering appropriate defect equilibria, we can see that an increased donor concentration leads to an increased concentration of barium vacancies. A similar effect is obtained by an increase in the oxygen partial pressure during the sintering and by an increase of the sintering temperature. Hence Hanke and Schmelz suggest that it is the barium vacancies which stop the growth of the $BaTiO_3$ grains after they increase in number to a certain limiting level. This limiting number does not depend on the kind of donor additive. The presence of acceptor additives, a reduction in the oxygen partial pressure, or an increase in the sintering temperature, on the other hand, decrease the concentration of these vacancies. These conclusions have been proved by calculations made by these investigators. Their model does not, however, explain fully how the barium vacancies inhibit the growth of $BaTiO_3$ grains.

4.4.4 Formation of potential barriers on the $BaTiO_3$ surface

The possibility of the formation of potential barriers on the surface of $BaTiO_3$ seems to be interesting from the point of view of the manufacture of capacitive devices. Heywang (1971) gives the results of studies on the behaviour of barium titanate junctions with various metals (Fig. 4.26). We can see from the figure that in the junctions with metals that do not reduce surface Ti^{+4} ions the surface resistivity is high, in contrast to the junctions with less noble metals. The capacitors with surface potential barriers have, however, a very low conduction voltage; much better parameters are obtained when the capacitors are prepared by reducing entire $BaTiO_3$ grains

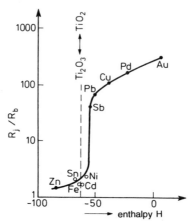

Fig. 4.26. The ratio of the resistivity R_j of the junction to the resistivity R_b of the bulk of BaTiO$_3$ grains when the electrodes are made of various metals. The dashed line indicates the enthalpy of the reduction of Ti^{+4} ions (after Heywang, 1971; reprinted by permission of Chapman & Hall)

and, then, oxidizing their surface layers. This technology will be discussed in Chapter 5.

4.4.5 Potential barriers at the boundaries of BaTiO$_3$ grains

The formation of appropriate potential barriers at the boundaries of the BaTiO$_3$ grains is the chief objective in manufacturing PTCR materials. At the beginning of the present chapter, we discussed the defect structure of the interior of these grains and described the models that represent the formation of acceptor states associated with barium vacancies. Many investigators consider the acceptor states to be responsible for the formation of the potential barriers. According to Heywang (1961, 1971) model, the acceptor states on the BaTiO$_3$ surfaces form as a result of the adsorption of oxygen there during the sintering process, which leads to the formation of oxygen vacancies.

In order to increase the potential barrier, the material is additionally doped with metal oxides that are placed on the surfaces of the BaTiO$_3$ grains. An example of such a dopant may be manganese oxide. Manganese is able to change its valency in the oxide reversibly according to

$$\text{MnO}_2 \underset{}{\overset{870 \text{ K}}{\rightleftarrows}} \text{Mn}_2\text{O}_3 \underset{}{\overset{1170 \text{ K}}{\rightleftarrows}} \text{Mn}_3\text{O}_4 \qquad (4.11)$$

In BaTiO$_3$ doped with La^{+3} ions, a number of additional Mn^{+3} ions compensate the equal number of lanthanum ions. Mn^{+2} ions may be expected to act as donors, whereas Mn^{+4} (empty Mn^{+3} states) will act as acceptors. At higher temperatures, we have a mixture of bi- and trivalent ions, but during cooling the equilibrium shifts towards the mixture of Mn^{+3} and Mn^{+4} (Lambeck and Jonker, 1978). The Mn^{+4} ions form acceptor states on the surfaces of the crystallites, and similar states in the vitreous intergranular phase are reported by these investigators. According to what has been said earlier, the presence of these states increases the height of the potential barrier. This hypothesis has been confirmed by the fact that, when the samples prepared by Jonker were additionally doped with manganese oxide (0.1 at%), the potential barriers formed on their surfaces were higher.

A similar effect is obtained by doping $BaTiO_3$ with oxides of other transition metals, such as Cu, Fe, Cr, V and Ru. Whereas in conventional PTCR thermistors, the abrupt change of resistivity corresponds to two to four orders of magnitude, the thermistors doped with these metal oxides exhibit a resistivity change by as much as seven orders of magnitude.

Wang and Umeya (1991) have measured the ac electrical properties of PTCR thermistors and studied the trap states introduced at grain boundaries when manganese oxide is added to the ceramics. They found that Mn increases the density of trap states positioned 1.35 eV below the conduction band. The trapped charge in samples without Mn was only found at about 0.35 eV below the conduction band. Hence, the trap states introduced by manganese are believed to enhance substantially the PTCR effect.

As in the case of ZnO varistors, chemisorbed oxygen is considered to be here a potential source of acceptor states at the grain boundaries. The chemisorbed oxygen captures the conduction electrons from the bulk, thereby enhancing the formation of the Schottky barrier (Nasrallah et al., 1984). Hasegawa et al. (1991) have recently demonstrated the enhanced grain boundary diffusion of O^{18} isotope in PTCR ceramics. The increased O^{18} content at the grain boundaries, measured using secondary ion mass spectrometry (SIMS), was correlated with the enhancement of the PTCR effect. Ho and Fu (1992) report an increase of the magnitude of the PTCR effect during reoxidation of the ceramics sintered and cooled in a reducing atmosphere. The reduced material already possessed inherent acceptor states and their density greatly increased after treatment in oxygen. Other possible chemisorbed species that give rise to a strong PTCR effect are fluorine ions. Alles et al. (1989) observed a PTCR jump by more than three orders of magnitude in fluorinated samples made of undoped, atmospherically reduced $BaTiO_3$. The PTCR action did not decrease after further treatment in N_2 and CO, which suggests that fluorine is bound more strongly than chemisorbed oxygen.

Chiang and Takagi (1990b) proposed another explanation of the origins of the electrical barriers formed at grain boundaries. When examining a commercial PTCR resistor by scanning transmission electron microscopy, they found that the oxygen content at the grain boundaries was slightly increased as compared to that observed in quenched samples. Based on the space charge segregation theory (briefly outlined in Section 2.3), they suggest that it is the high-temperature defect segregation which is responsible for the formation of acceptor states at grain boundaries. These acceptor defects are believed to be barium vacancies. In practical devices, the PTCR effect is enhanced by intentional doping with additional acceptors, such as Mn or chemisorbed oxygen.

Desu and Payne (1990a–d, 1992) suggest that n–i–n (semiconductor–insulator–semiconductor) layers exist at the grain boundaries, and that they account for the PTCR effect. These layers form due to the donor and/or acceptor segregation. The occurrence of the donor segregation was questioned by Chiang and Takagi (1990a, 1992), especially in the case of Nb and Sb whose ionic radii are comparable to that of Ti. The role of the interfacial segregation in electronic ceramics is still not clear and needs further investigation. Modern experimental techniques and instruments allow us to observe quantitatively the near-atomic scale processes, but to prepare the sample in a non-destructive way is always a problem. The relation between certain properties of the sample observed microscopically and the behaviour of the real device in macroscopic terms is often difficult to assess.

4.4.6 Two-stage PTCR effect

During experiments with PTCR materials doped with surface-active acceptor additives, such as CuO, aimed at increasing the resistivity jump, a two-stage PTCR effect has been observed as shown in Fig. 4.27 (Heywang and Wersing, 1974; Kuwabara and Yanagida, 1972, 1973; Kuwabara, 1983d). The additional maximum of resistivity occurs at a temperature below the Curie point. This effect has been observed in materials doped with small (0.01 wt%) amounts of CuO. As can be seen from the figure, with increasing CuO content, the temperature coefficient of resistivity of the semiconducting material becomes negative (the material transforms into NTCR material). Heywang and Wersing (1974) explained this behaviour in terms of the dielectric mechanism. They only considered those portions of the grain boundary where the conditions for the flow of electric current are most favourable, i.e., where the potential barrier is the lowest. This, in turn, occurs in regions where the surface charge is compensated most strongly due to the spontaneous polarization that takes place below the Curie temperature.

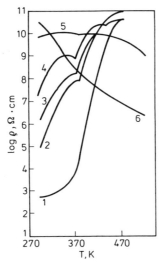

Fig. 4.27. The resistivity-temperature curves of CuO-doped BaTiO$_3$; the numbers 1 to 6 denote curves corresponding to various CuO contents: 1—the smallest concentration, and 6—the greatest concentration (after Kuwabara, 1983d; reprinted by permission of Chapman & Hall)

Equation (4.3) may be put in the form

$$eV_b = \frac{e^2 N_D b^2}{2\varepsilon\varepsilon_0} \qquad (4.12)$$

where $b = N_S^-/N_D$, and N_S^- is the concentration of the fully occupied acceptor levels.

For the spontaneous polarization P_S, b was replaced by b' defined as

$$b' = \frac{1}{N_D}\left(N_S^- - \frac{P_S}{e}\right) \qquad (4.13)$$

This equation was then substituted into equation (4.12). The plots of eV_b/kT versus T calculated numerically appeared to be consistent with experimental curves. The conductivity variation with acceptor concentration, i.e., with the additive (CuO in this case) concentration, was analogous to those shown in Fig. 4.27. Heywang and Wersing's reasoning confirmed, once again, the validity of the PTCR model described earlier.

4.4.7 Conduction mechanism above the Curie temperature

By analysing the current-voltage characteristics of PTCR resistors, Kuwabara (1983a) considered the conduction mechanism that operates in PTCR resistors above the Curie temperature. Figure 4.28 shows the current-voltage characteristics obtained by this investigator at various temperatures. These

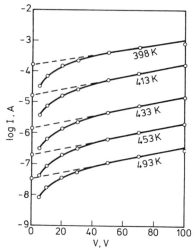

Fig. 4.28. Current–voltage characteristics of PTCR resistors measured at various temperatures (after Kuwabara, 1983a; reprinted by permission of the American Ceramic Society)

curves may suggest that the flowing current is thermally activated (i.e., a diffusion current) and given by

$$I = I_0 \exp\left(\frac{eV}{nkT}\right) \qquad (4.14)$$

The value of the coefficient n provides information about whether the current is purely diffusive ($n = 1$) or is due also to the generation and recombination of hole–electron pairs ($n = 2$). Equation (4.14) applies to a single p–n junction and, thus, when dealing with a ceramic material, the statistical $BaTiO_3$ grain-size distribution should be assumed in the calculations. Kuwabara obtained $n = 1.1$ for the material with an average grain size of 2 µm, and $n = 2.8$ when the average grain size was 5 µm. These results also confirm the validity of the Heywang model.

Kuwabara (1984) also calculated the height of the potential barrier at the $BaTiO_3$ grain boundary, based on the measured values of the resistivity and dielectric constant as functions of temperature. According to Heywang, the resistivity of a potential barrier is given by

$$\varrho = \varrho_0 \exp\left(\frac{eV_b}{kT}\right) = \varrho_0 \exp\left(\frac{e^2 N_S^2}{2\varepsilon\varepsilon_0 N_D kT}\right) \qquad (4.15)$$

where $\varrho_0 = $ const. Knowing the capacitance C of the specimen at a frequency of 1 kHz, Kuwabara calculated the dielectric constant $\bar{\varepsilon}$ from the equation: $\bar{\varepsilon} = Cd/\varepsilon_0 S_0$, where d is the thickness of the specimen, and S_0 is the surface area of the electrodes. The relationship between ε and $\bar{\varepsilon}$ is given by $\bar{\varepsilon} = r\varepsilon/2b$, where r is the average diameter of the BaTiO$_3$ grain, and $b = N_S/N_D$. Equation (4.15) may be rearranged to take the form in which $\bar{\varepsilon}$ acquires a macroscopic value:

$$\varrho = \varrho_0 \exp\left(\frac{e^2 N_S r}{\bar{\varepsilon}\varepsilon_0 kT}\right) \quad (4.16)$$

Since the material examined was highly porous, the effective active cross-section of the intergranular contacts, S, should be inserted. This cross-section is related to the surface area of the electrodes by the function $f(p) = S/S_0$. When this is done, we obtain

$$\varrho = \varrho_0 \exp\left(\frac{e^2 N_S r f(p)}{4\bar{\varepsilon}\varepsilon_0 kT}\right) \quad (4.17)$$

The variation of $\ln \varrho$ with $1/\bar{\varepsilon}T$, obtained experimentally, appeared to be linear. Hence by assuming that N_S, r and $f(p)$ are independent of temperature, Kuwabara calculated the heights of the potential barrier at the Curie temperature and at a temperature at which the resistivity of the material was a maximum. Taking $r = 2$ μm and $r = 5$ μm, he obtained the barrier heights of 0.06–0.09 eV and 0.66–0.69 eV, respectively. In some samples these heights were even 0.11 and 0.88 eV. The fact that the plot of $\varrho(1/\bar{\varepsilon}T)$ was linear also confirmed the validity of the Heywang (1961, 1971) model.

REFERENCES

Al-Allak H. M., Parry T. V., Russell G. J. and Woods J. (1988), Effects of aluminium on the electrical and mechanical properties of PTCR BaTiO$_3$ ceramics as a function of the sintering temperature, *J. Mat. Sci.*, **23** (3), 1083.

Alles A. B., Amarakoon V. R. W. and Burdick V. L. (1989), Positive temperature-coefficient of resistivity effect in undoped, atmospherically reduced barium-titanate, *J. Am. Ceram. Soc.*, **72** (1), 148.

Basu R. N. and Maiti H. S. (1987), Study of the semiconducting barium titanate by impedance measurement, in: Vincenzini (ed.) (1987), *High-Tech Ceramics*, Elsevier, Amsterdam, 1883.

Berglund C. N. and Baer W. S. (1967), Electron transport in single-domain ferroelectric barium titanate, *Phys. Rev.*, **157** (2), 358.

Brauer H. (1974), Resistance anomaly in semiconductor BaTiO$_3$ ceramics in range below Curie point, *Solid State Electron.*, **17** (10), 1013.

Casella R. C. and Keller S. P. (1959), Polarized light transmission of BaTiO$_3$ single crystals, *Phys. Rev.*, **116** (6), 1469.

Chaput F. and Boilot J. P. (1987), Chemically derived barium titanate gels and ceramics, in: Vincenzini (ed.) (1987), *High-Tech Ceramics*, Elsevier, Amsterdam, 1459.

Chiang Y.-M. and Takagi T. (1990a), Grain-boundary chemistry of barium-titanate and strontium-titanate: I, High-temperature equilibrium space-charge, *J. Am. Ceram. Soc.*, **73** (11), 3278.

Chiang Y.-M. and Takagi T. (1990b), Grain-boundary chemistry of barium-titanate and strontium-titanate: II, Origin of electrical barriers in positive-temperature-coefficient thermistors, *J. Am. Ceram. Soc.*, **73** (11), 3286.

Chiang Y.-M. and Takagi T. (1992), Comment on "Interfacial segregation in perovskites: I–IV", *J. Am. Ceram. Soc.*, **75** (7), 2017.

Chin T. S. and Lou C. H. (1991), Porous single-phase $PbTiO_3$ as PTCR ceramics, *Mater. Lett.*, **11** (10–12), 379.

Daniels J., Hardtl K. H. and Wernicke R. (1978/79), The PTC effect of barium titanate, *Philips Tech. Rev.*, **38** (3), 73.

Desu S. B. and Payne D. A. (1990a), Interfacial segregation in perovskites: I, Theory, *J. Am. Ceram. Soc.*, **73** (11), 3391.

Desu S. B. and Payne D. A. (1990b), Interfacial segregation in perovskites: II, Experimental evidence, *J. Am. Ceram. Soc.*, **73** (11), 3398.

Desu S. B. and Payne D. A. (1990c), Interfacial segregation in perovskites: III, Microstructure and electrical properties, *J. Am. Ceram. Soc.*, **73** (11), 3407.

Desu S. B. and Payne D. A. (1990d), Interfacial segregation in perovskites: IV, Internal boundary layer devices, *J. Am. Ceram. Soc.*, **73** (11), 3416.

Desu S. B. and Payne D. A. (1992), Reply to "Comment on 'Interfacial segregation in perovskites: I–IV'", *J. Am. Ceram. Soc.*, **75** (7), 2020.

Drofenik M. (1987), Oxygen partial pressure and grain growth in donor-doped $BaTiO_3$, *J. Am. Ceram. Soc.*, **70** (5), 311.

Eberspachel O. (1962), Einbau von Antimon in Bariumtitanat $BaTiO_3$, *Naturwissen*, **49** (7), 155.

Eibl O., Pongratz P., Skalicky P. and Schmelz H. (1987), Formation of (111) twins in $BaTiO_3$ ceramics, *J. Am. Ceram. Soc.*, **70** (8), C-195.

Gerthsen P. and Hardtl K. H. (1963), Eine Methode zum direkten Nachweis von Leitfähigkeitsinhomogenitäten an Korngrenzen, *Z. Naturforsch.*, **18A**, 423.

Gerthsen P., Groth R., Hardtl K. H., Heese D. and Reik H. G. (1965a), The small polaron problem and optical effects in barium titanate, *Solid State Commun.*, **3** (1), 165.

Gerthsen P., Groth R. and Hardtl K. H. (1965b), Halbleitereigenschaften des $BaTiO_3$ im Polaronenbild, *Phys. stat. solidi*, **11**, 303.

Gerthsen P. and Hoffmann B. (1973), Current-voltage characteristics and capacitance of single grain boundaries in semiconducting $BaTiO_3$ ceramics, *Solid State Electron.*, **16** (5), 617.

Haanstra H. B. and Ihrig H. (1977), Voltage contrast imaging of PTC-type $BaTiO_3$ ceramics having low and high titanium excess, *Phys. stat. solidi (a)*, **39** (1), K7.

Haanstra H. B. and Ihrig H. (1980), Transmission electron microscopy at grain boundaries of PTC-type $BaTiO_3$ ceramics, *J. Am. Ceram. Soc.*, **63** (5–6), 288.

Haayman P. W., Dam R. W. and Klasens H. A. (1955), Method of preparation of semiconducting materials, German Patent 929,350, June 23.

Hanke L. (1979), Theorie der Sperrschichteffekte in halbleitender Bariumtitanat-Keramik, *Siemens Forsch. -u. Entwickl. Ber.*, **8** (4), 209.

Hanke L. and Schmelz H. (1982), The significance of barium vacancies with regard to the doping anomaly of barium titanate ceramics, *Ber. Dtsch. Keram. Ges.*, **59** (4), 1.

Hasegawa A., Fujitsu S., Koumoto K. and Yanagida H. (1991), The enhanced penetration of oxygen along the grain-boundary in semiconducting barium-titanate, *Jpn. J. Appl. Phys.*, **30** (6), 1252.

Hennings D. F. K., Janssen R. and Reynen P. J. L. (1987), Control of liquid-phase-enhanced discontinuous grain growth in barium titanate, *J. Am. Ceram. Soc.*, **70** (1), 23.

References

Heywang W. (1961), Bariumtitanat als Sperrschichthalbleiter, *Solid State Electron.*, **3** (1), 51.

Heywang W. (1964), Resistivity anomaly in doped barium titanate, *J. Am. Ceram. Soc.*, **47** (10), 484.

Heywang W. (1970), The Curie shift in $BaTiO_3$ by hetero-substitution, *Ferroelectrics*, **1** (1), 177.

Heywang W. (1971), Semiconducting barium titanate, *J. Mat. Sci.*, **6** (9), 1214.

Heywang W. and Wersing W. (1974), Anomalous temperature resistance characteristic of highly acceptor doped $BaTiO_3$ PTC, *Ferroelectrics*, **7**, 361.

Heywang W., Schumacher K. and Thomann H. (1976), Ferroelektrische Keramik in der Elektroindustrie, *Ber. Dtsch. Keram. Ges.*, **53** (10), 358.

Ho I. C. and Fu S. L. (1992), Effect of reoxidation on the grain-boundary acceptor-state density of reduced $BaTiO_3$ ceramics, *J. Am. Ceram. Soc.*, **75** (3), 728.

Hoffmann B. (1973), Model for grain-boundary resistance in doped $BaTiO_3$ ceramics, *Solid State Electron.*, **16** (5), 623.

Howng Wei-Yean and McCutcheon C. (1983), Electrical properties of semiconducting $BaTiO_3$ by liquid-phase sintering, *Am. Ceram. Soc. Bull.*, **62** (2), 231.

Ihrig H. and Puschert W. (1977), A systematic experimental and theoretical investigation of the grain-boundary resistivities of n-doped $BaTiO_3$ ceramics, *J. Appl. Phys.*, **48** (7), 3081.

Ihrig H. (1978), The PTC effect of semiconducting $BaTiO_3$ ceramics as a function of the titanium excess, *Ber. Dtsch. Keram. Ges.*, **55** (6), 319.

Ihrig H. and Klerk M. (1979), Visualization of the grain-boundary potential barriers of PTC-type $BaTiO_3$ ceramics by cathodoluminescence in an electron probe microanalyzer, *Appl. Phys. Lett.*, **35** (4), 307.

Ihrig H. (1983), Physics and technology of PTC-type $BaTiO_3$ ceramics, in: Yan M. F. and Heuer A. H. (eds.) (1983), *Additives and Interfaces in Electronic Ceramics. Advances in Ceramics*, Vol. 7, Am. Ceram. Soc., Columbus, Ohio, 117.

Inaba M., Miyayama M. and Yanagida H. (1992), Infrared sensing properties of positive temperature-coefficient thermistors with large temperature coefficients of resistivity, *J. Mat. Sci.*, **27** (1), 127.

Janitzki A. S., Hoffmann B. and Gerthsen P. (1979), Sintering behavior of doped-undoped ceramics $BaTiO_3$ composites, *J. Am. Ceram. Soc.*, **62** (7–8), 422.

Jonker G. H. (1964), Some aspects of semiconducting barium titanate, *Solid State Electron.*, **7** (12), 895.

Jonker G. H. (1981), Equilibrium barriers in PTC thermistors, in: Levinson L. M. and Hill D. C. (eds.) (1981), *Grain Boundary Phenomena in Electronic Ceramics. Advances in Ceramics*, Vol. 1, Am. Ceram. Soc., Columbus, Ohio, 155.

Ki Hyun Yoon and Eun Hong Lee (1987), PTCR effects in semiconductive $BaTiO_3$ prepared in molten KCl, in: Vincenzini (ed.) (1987), *High-Tech Ceramics*, Elsevier, Amsterdam, 1873.

Kulwicki B. M. and Purdes A. J. (1970), Diffusion potentials in $BaTiO_3$ and the theory of PTC materials, *Ferroelectrics*, **1** (1), 253.

Kulwicki B. M. (1981), PTC materials technology, 1955–1980, in: Levinson L. M. and Hill D. C. (eds.) (1981), *Grain Boundary Phenomena in Electronic Ceramics. Advances in Ceramics*, Vol. 1, Am. Ceram. Soc., Columbus, Ohio, 138.

Kuwabara M. and Yanagida H. (1972), Two-step anomalous increase of resistivity on La-doped $BaTiO_3$–Bi_2O_3 composite ceramics with surface barrier layer, *Jpn. J. Appl. Phys.*, **11** (8), 1130.

Kuwabara M. and Yanagida H. (1973), On the surface structure of the La-doped BaTiO$_3$–Bi$_2$O$_3$ composite ceramics showing a two-step anomalous increase of resistivity, *Jpn. J. Appl. Phys.*, **12** (9), 1351.

Kuwabara M. and Yanagida H. (1980), Microstructure and the PTCR effect of semiconducting barium titanate ceramics, *J. Phys. Soc. Jpn.*, **49** (Suppl. B), 146.

Kuwabara M. (1981a), Fabrications of porous PTCR thermistors and their PTCR effects, *Jpn. J. Appl. Phys.*, **20** (Suppl. 4), 131.

Kuwabara M. (1981b), Influence of stoichiometry on the PTCR effect in porous barium titanate ceramics, *J. Am. Ceram. Soc.*, **64** (12), C-170.

Kuwabara M. (1981c), Effect of microstructure on PTCR effect in semiconducting barium titanate ceramics, *J. Am. Ceram. Soc.*, **64** (11), 639.

Kuwabara M. (1983a), Explanation for the PTCR effect in barium titanate ceramics, in: Yan M. F. and Heuer A. H. (eds.) (1983), *Additives and Interfaces in Electronic Ceramics. Advances in Ceramics*, Vol. 7, Am. Ceram. Soc., Columbus, Ohio, 128.

Kuwabara M. (1983b), PTCR effect in barium titanate ceramics, in: Yan M. F. and Heuer A. H. (eds.) (1983), *Additives and Interfaces in Electronic Ceramics. Advances in Ceramics*, Vol. 7, Am. Ceram. Soc., Columbus, Ohio, 137.

Kuwabara M. and Inoue H. (1983c), Effect of ambient gas on the electrical conduction in semiconducting barium titanate ceramics, in: Seiyama T., Fueki K., Shiokawa J. and Suzuki S. (eds.) (1983), *Chemical Sensors, Anal. Chem. Symposia Series*, Vol. 17, Elsevier, Amsterdam, 182.

Kuwabara M. (1983d), Effect of CuO addition on the electrical and dielectrical properties of semiconducting barium titanate ceramics, *J. Mat. Sci. Lett.*, **2** (8), 403.

Kuwabara M. (1983e), PTCR effect in barium titanate powder compacts mixed with ohmic-contact metals, *Am. Ceram. Soc. Bull.*, **62** (6), 708.

Kuwabara M. and Kumamoto K. (1983f), PTCR characteristics in barium titanate ceramics with Curie points between 60° and 360°C, *J. Am. Ceram. Soc.*, **66** (11), C-214.

Kuwabara M. (1984), Determination of the potential barrier height in barium titanate ceramics, *Solid State Electron.*, **27** (11), 929.

Kuwabara M., Suemura S. and Kawahara M. (1985), Preparation of high-Curie-point barium-lead titanates and their PTCR characteristics, *Am. Ceram. Soc. Bull.*, **64** (10), 1394.

Kuwabara M., Nakao K. and Okazaki K. (1988), Instability of the characteristics of the positive temperature coefficient of resistivity in high-Curie-point barium-lead titanate ceramics and their grain structures, *J. Am. Ceram. Soc.*, **71** (2), C-110.

Kuwabara M. (1990), Lead titanate ceramics with positive temperature coefficient of resistivity, *J. Am. Ceram. Soc.*, **73** (5), 1438.

Lambeck P. V. and Jonker G. H. (1978), Ferroelectric domain stabilization in BaTiO$_3$ by bulk ordering of defects, Ferroelectrics, **22** (1/2), 729.

Levinson L. M. and Hill D. C. (eds.) (1981), *Grain Boundary Phenomena in Electronic Ceramics. Advances in Ceramics*, Vol. 1, Am. Ceram. Soc., Columbus, Ohio.

Lin T. F., Hu C. T. and Lin I. N. (1990), Influence of stoichiometry on the microstructure and positive temperature-coefficient of resistivity of semiconducting barium-titanate ceramics, *J. Am. Ceram. Soc.*, **73** (3), 531.

Macdonald J. R. (ed.) (1987), *Impedance Spectroscopy*, John Wiley & Sons, New York.

Matsuo Y. and Sasaki H. (1971), Exaggerated grain growth in liquid-phase sintering of BaTiO$_3$, *J. Am. Ceram. Soc.*, **54** (9), 471.

Mostaghaci H. and Brook R. J. (1983), Production of dense and fine grain size BaTiO$_3$ by fast firing, *Trans. J. Br. Ceram. Soc.*, **82** (5), 167.

References

Nasrallah M. M., Anderson H. V., Agarwal A. K. and Flandermeyer B. (1984), Oxygen activity dependence of the defect structure of La-doped $BaTiO_3$, *J. Mat. Sci.*, **19** (10), 3159.

Nemoto H. and Oda T. (1980), Direct examinations of PTC action of single grain boundaries in semiconducting $BaTiO_3$ ceramics, *J. Am. Ceram. Soc.*, **63** (7–8), 398.

Nemoto H. and Oda T. (1981), Direct examinations of electrical properties of single grain boundaries in $BaTiO_3$ PTC ceramics, in: Levinson L. M. and Hill D. C. (eds.) (1981), *Grain Boundary Phenomena in Electronic Ceramics. Advances in Ceramics*, Vol. 1, Am. Ceram. Soc., Columbus, Ohio, 167.

Philips Data Handbook (1984), Components and Materials, Part 11.

Phule P. P., Raghavan S. and Risbud S. H. (1987), Comparison of $Ba(OH)_2$, BaO and Ba as starting materials for the synthesis of barium titanate by the alkoxide method, *J. Am. Ceram. Soc.*, **70** (5), C-108.

Rase D. E. and Roy R. (1955a), Phase equilibria in the system $BaO-TiO_2$, *J. Am. Ceram. Soc.*, **38** (3), 102.

Rase D. E. and Roy R. (1955b), Phase equilibria in the system $BaTiO_3-SiO_2$, *J. Am. Ceram. Soc.*, **38** (11), 389.

Rehme H. (1968), Elektronenmikroskopische Untersuchung zum Mechanismus von Bariumtitanat Kaltleiter Keramik, *Phys. stat. solidi*, **26** (1), K1.

Saburi O. (1961), Semiconducting bodies in the family of barium titanate, *J. Am. Ceram. Soc.*, **44** (2), 54.

Schmelz H. (1969), Incorporation of antimony into the barium titanate lattice, *Phys. stat. solidi*, **31** (1), 121.

Schmelz H. and Meyer A. (1982), The evidence for anomalous grain growth below the eutectic temperature in $BaTiO_3$ ceramics, *Ber. Dtsch. Keram. Ges.*, **59** (8–9), 436.

Tseng T. Y. and Wang S. H. (1990), Ac electrical-properties of high-Curie-point barium lead titanate PTCR ceramics, *Mater. Lett.*, **9** (4), 164.

Ueoka H. and Yodogawa M. (1974), Ceramic manufacturing technology for the high performance PTC thermistor, *IEEE Trans. Manuf. Tech.*, **MFT-3** (2), 77.

Vincenzini (ed.) (1987), *High-Tech Ceramics*, Elsevier, Amsterdam.

Wang D. Y. and Umeya K. (1991), Spontaneous polarization screening effect and trap-state density at grain-boundaries of semiconducting barium-titanate ceramics, *J. Am. Ceram. Soc.*, **74** (2), 280.

Wang S.-H., Hwang F. S. and Tseng T. Y. (1990), Fabrication of high-Curie-point barium-lead titanate PTCR ceramics, *J. Am. Ceram. Soc.*, **73** (9), 2767.

Yan M. F. and Heuer A. H. (eds.) (1983), *Additives and Interfaces in Electronic Ceramics. Advances in Ceramics*, Vol. 7, Am. Ceram. Soc., Columbus, Ohio.

Yoneda Y., Kato H. and Sasaki H. (1976), Sintering process of semiconductive $BaTiO_3$, ceramics, *J. Am. Ceram. Soc.*, **59** (11–12), 531.

Young Ho Han, Appleby J. B. and Smyth D. M. (1987), Calcium as an acceptor impurity in $BaTiO_3$, *J. Am. Ceram. Soc.*, **70** (2), 96.

Young-Sung Yoo, Jeong-Joo Kim and Doh-Yeon Kim (1987), Effect of heating rate on the microstructural evolution during sintering of $BaTiO_3$ ceramics, *J. Am. Ceram. Soc.*, **70** (11), C-322.

CHAPTER 5

Boundary Layer Capacitors

Since capacitors belong to the group of most widely used discrete electronic components, continuous efforts have been made to improve their parameters and to minimize their size. Studies on PTCR thermistors, carried out for many years, indicated that the ceramic materials of which they are manufactured may also be used for fabricating ceramic capacitors. Since that time, new ceramic materials classified in a separate group have been developed, specially intended for the manufacture of capacitors.

In the first part of this chapter we describe how the technology of these new GBBL (grain boundary barrier layer) materials has been developed and then we discuss the recent results of studies on their properties. An extensive review on that matter was given by Goodman (1981).

Daniels, Hardtl and Wernicke (1978/79) have described in detail the operation mechanism of PTCR resistors (cf. Section 4.4). They describe how the properties of ceramics depend on the rate at which they are cooled from the sintering temperature. Let us remember that the defect structures of the near-boundary regions and of the interior of La-doped $BaTiO_3$ grains, established during the cooling cycle, differ from one another. The diffusion rate of the V_{Ba} vacancies (singly or doubly ionized) in the $BaTiO_3$ crystal lattice is relatively low. During cooling, they diffuse towards the centre of the grain, compensating the donor impurities, such as La. As a result, a layered structure forms, in which the compensated layer has a width l_D that increases with decreasing cooling rate (Fig. 4.24). The cooling rate also affects the shape of the potential barriers established at the grain boundaries as shown in Fig. 5.1.

When we allow the insulating near-boundary layers to grow to an appropriate width, the material that we obtain will have a heterogeneous structure with semiconducting grain bulk. This material is very well suited for

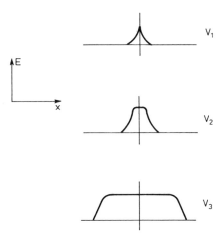

Fig. 5.1. Diagram showing how the shape of the potential barrier at the BaTiO$_3$ grain boundary varies as a function of the rate of cooling (V) from the sintering temperature ($V_1 < V_2 < V_3$); redrawn from Daniels et al. (1978/79)

manufacturing GBBL capacitors. The effective dielectric constant of the capacitors fabricated of this material may even exceed 300 000, thanks to the fact that its individual grains function as microcapacitors connected, at the same time, in series and in parallel. Figure 5.2 shows various types of boundary-layer capacitors.

The transition from a PTCR resistor to a GBBL capacitor, as described above, was directly observed by Sundaram (1990), who examined it using ac complex impedance spectroscopy. The thickness of the insulating layer at the grain boundaries of Sb-doped BaTiO$_3$ ceramics was simply increased by increasing the sintering time. In samples sintered for a shorter time (up to 20 hours), the PTCR effect occurred, but the room-temperature resistivity increased continuously. At longer sintering times (30 hours or more), the dc resistance exceeded 1 MΩ, and the dielectric loss tangent ($\tan \delta$) and the dielectric constant decreased. Scanning electron microscopy examinations showed that no additional grain growth occurred during the prolonged sintering. The width of the grain boundaries apparently increased.

Because of their nature, GBBL capacitors may only be used in low-voltage circuits (Yamaoka et al., 1983). This is, however, consistent with the requirements of modern electronics which are based on semiconducting components. The prototypes of GBBL capacitors were the so-called surface-barrier capacitors fabricated in 1948 (Roup and Butler, 1950). They were composed of two layers: conducting and insulating. They were fabricated by sintering (Ba, Sr)TiO$_3$ doped (0.5 wt%) with an oxide mixture of the composition (30% Nd + 20% Pr + 45% La + 5% Sm) in an oxidizing atmosphere (in order to make it insulating). Then, after covering the front faces of the sample

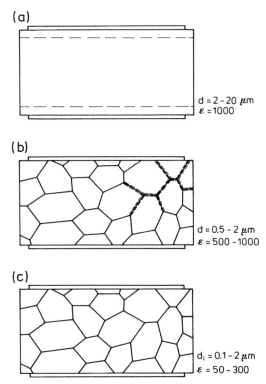

Fig. 5.2. Boundary-layer capacitors of various type: (a) surface layers, (b) grain-boundary layers formed by defect diffusion, (c) insulating layers grown due to the presence of the intergranular phase. Typical thicknesses d and dielectric constants ε are also shown. Reprinted from Wernicke (1981a) by permission of the American Ceramic Society

with carbon which acted as a reducing agent, the ceramic was refired at a lower temperature. The reverse procedure was also employed: the material was first sintered in a reducing atmosphere and then refired in air.

In 1957, Glaister (1962) obtained a patent for the technology of GBBL capacitors. He sintered high-purity $BaTiO_3$ in hydrogen at a temperature of 1720 K; then the material was quickly cooled to 1270 K and, then, slowly cooled to room temperature. The porous material, with a grain size of about 1 µm, obtained in this way was next oxidized (at its grain boundaries) at a temperature between 920 and 1270 K. This technology is still in general use. Its basic stages include:

(1) sintering the material in air until an insulating material of high density is obtained;

(2) reducing the material throughout its entire volume by firing in hydrogen or in another reducing atmosphere;

(3) reoxidizing the insulating layers at the grain boundaries;
(4) depositing the electrodes.

These technological processes have been optimized to obtain boundary layers as thin as possible and sharply separated from the conducting regions. Next, Waku (1967, 1970) improved the process by introducing acceptor additives, such as Mn and Cu, on the oxidized surface of the surface-barrier capacitors. With this technique, the insulating layer forms not only as a result of firing the material in an oxidizing atmosphere but also due to the action of the acceptors.

Brauer and Kuschke (1971, 1974) were first to describe a single-stage sintering GBBL manufacturing technique, in which the insulating layer was formed as a result of the strong action of acceptors (CuO) and donors (Sb_2O_3 or La_2O_3) introduced in combination. The technique appeared to be successful if the process conditions permitted the $BaTiO_3$ grains to grow to an appropriate diameter (100–300 μm). Then Waku et al. (1971) proposed a similar technology for another material, namely a mixture of the composition $Ba(Ti_{0.9}Sn_{0.1})O_3$ added with 0.1 mol% Dy_2O_3, 0.4 wt% SiO_2, 0.5 mol% $BaCO_3$ and 0.6 mol% CuO. The mixture was sintered in nitrogen and then cooled in air. In the meantime, a need arose for capacitors that operate at very high frequencies, of the order of GHz. It follows from theory that these capacitors maintain large capacitance values at high frequencies, when their grain bulks show high conductivity. The material that satisfies this requirement appeared to be strontium titanate, $SrTiO_3$, in which the electron mobility is greater than that in barium titanate by a factor of ten. Waku et al. (1970) described how to fabricate boundary layer capacitors using the compound $(Ba,Sr)TiO_3$ doped with Dy_2O_3. The compound was sintered in a reducing atmosphere (a thin ceramic foil was covered, on both sides, with CuO layers). Although studies on this technology were carried out for a long time, mass production of the $SrTiO_3$-based GBBL capacitors has only recently been started; at the present, $SrTiO_3$ is the basic material used for this production. The problems to overcome included the high dielectric losses, the strong variation of the electrical parameters with temperature, and the difficulties in choosing appropriate electrodes. Yamaoka and Matsui (1981) reported that GBBL capacitors (type II) with a dielectric constant of the order of 40 000–60 000, a dielectric loss factor $\tan \delta = 0.005$, a resistance greater than 3500 MΩ at a dc voltage of 50 V, and the temperature coefficient (within the range from 270 to 350 K) of the order of several percent or less were manufactured commercially in Japan in 1980. Japanese manufacturers produced capacitors of various rated capacitances and various designs, such as those suitable to substitute organic dielectric capacitors.

Let us now discuss certain results of studies on the GBBL capacitor-type materials. Wernicke (1981a) describes the typical technique of fabricating SrTiO$_3$-based GBBL capacitors:

(1) Preparation of raw materials of typical composition: Sr(Ti$_{1-x}$Nb$_x$)O$_3$, with $x = 0.01$–0.05, are mixed with the following additives (the starting powders are mixed using conventional ceramic techniques):
— an excess of Ti (of the order of 1 wt%), which, during sintering, forms the intergranular phase,
— SiO$_2$, Al$_2$O$_3$, GeO$_2$, ZnO as the additives that improve the microstructure of the material. According to Klerk and Sanders (1981), these additives control the growth of the SrTiO$_3$ grains by affecting the melting temperature of the Ti phase. Al$_2$O$_3$ increases the melting point of this phase and hinders the grain growth, whereas SiO$_2$ has a converse effect. An addition of silica (about 1 wt%) can also increase the dielectric constant of the capacitor by a factor of two or more.

(2) Sintering at a temperature of 1670 to 1740 K in a reducing atmosphere (H$_2$/N$_2$, e.g., 5% H$_2$) in order to obtain coarse-grained ceramic material with a grain diameter of 20–50 µm and conductivity $\sigma_0 = 0.5$–2 $(\Omega \cdot \text{cm})^{-1}$. The intensive grain growth was achieved due to the presence of the liquid phase formed by the excess of Ti, and the high conductivity was due to doping with Nb or Ta.

(3) Depositing an oxide paste of typical composition: 50 wt% PbO + +45 wt% Bi$_2$O$_3$ + 5 wt% B$_2$O$_3$ mixed with appropriate organic substances.

(4) Refiring at a temperature between 1270 and 1470 K in a reducing atmosphere to enable the oxides to penetrate into the ceramic material and form the insulating layers.

(5) Depositing the electrodes, by e.g. evaporating CrNi or Au under vacuum or by firing an Ag paste.

Microscopic examinations made after the material has been refired clearly indicate the presence of another phase, 0.1–1 µm wide and composed of Pb, Bi, B and Ti, which surrounds the SrTiO$_3$ grains. At a temperature of about 870 K, as a result of various chemical reactions, the oxide layer deposited on the ceramic surface forms a vitreous coating which wets this surface very well. At about 1170 K, the oxides begin to penetrate, at a great speed, into the material; for example, they are fully distributed within a 10 mm thick sample after heating it at 1270 K for a few minutes. Figure 5.3 illustrates how the oxides penetrate into the ceramic material. At 1170 K, the surface film of the oxides begin to dissolve quickly the intergranular Ti phase. Since, at the highest temperature, the oxides in addition react with the grain surfaces, the insulating phase that is formed may be thicker than the Ti phase.

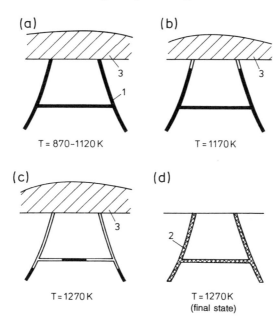

Fig. 5.3: (a)–(d) Subsequent stages of the penetration of oxides into the barium titanate grain boundaries after reacting with the intergranular TiO$_2$ phase; 1 — the intergranular phase (TiO$_2$) formed during sintering, 2 — the liquid phase formed as a result of the reaction between TiO$_2$ and the metal oxide mixture, 3 — oxide layers on the ceramic surface (the melting temperature is 870 K); reprinted from Wernicke (1981a) by permission of the American Ceramic Society

Stenton and Harmer (1983) examined the intergranular phase in SrTiO$_3$. They found that the layers of this phase were amorphous, except those positioned in Ti-rich regions.

Wernicke (1981b) explained why the dielectric constant of GBBL capacitors is so great. He assumed that the microstructure of the basic phase is an ideal arrangement of cubic grains with side d_g, just as was assumed in the first models describing the behaviour of ZnO varistors (Fig. 5.4). If we denote the thickness of the sample by d_c, the number of grain boundaries that contain an insulating layer of thickness d_i and dielectric constant ε_i positioned along this thickness is $n = d_c/d_g - 1 \approx d_c/d_g$. If now the total capacitance of

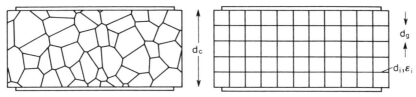

Fig. 5.4. Model microstructure of the GBBL-type ceramic material (after Wernicke, 1981b; reprinted by permission of the American Ceramic Society)

the sample is $C_{\text{tot}} = C_i/n$, then substituting $C_i = \varepsilon_0\varepsilon_i A/d_i$ (where A is the surface area of the electrodes) we obtain $C_{\text{tot}} = \varepsilon_0\varepsilon_i d_g A/d_i d_g$. Comparing this equation with the equation $C_{\text{tot}} = \varepsilon_0\varepsilon_{\text{eff}} A/d_c$ (where ε_{eff} is the effective dielectric constant), which describes the actual capacitance of the sample, we have

$$\varepsilon_{\text{eff}} = \varepsilon_i \frac{d_g}{d_i} \tag{5.1}$$

Thus, from the formal point of view, the GBBL capacitors may be described by the value of ε_{eff}, which has no direct physical significance. This criterion is valid if $d_c \gg d_g$. Computer calculations simulating the statistical distribution of the diameters and shapes of the grains that build up a ceramic material of homogeneous structure have shown that the error introduced by this approximation does not exceed 10%.

It follows from the preceding reasoning that the insulating layer formed at the SrTiO$_3$ grain boundary is not homogeneous. It is composed of the intergranular layer and two dielectric diffusion layers positioned on both its sides and formed as a result of the movement of Sr vacancies towards the interior of the grains, just as happens in PTCR resistors. Let us denote the properties of these two layers by the index 1 (σ_1, ε_1, d_1) and the properties of the intergranular layer by the index 2 (Fig. 5.5). Then we have

$$\frac{1}{\varepsilon_{\text{eff}}} = \frac{1}{d_g}\left(\frac{d_1}{\varepsilon_1} + \frac{d_2}{\varepsilon_2}\right) \tag{5.2}$$

During the formation of the intergranular layer, the mass of the sample in-

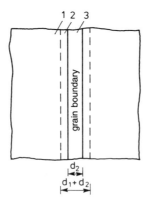

Fig. 5.5. Schematic representation of the boundary between semiconducting SrTiO$_3$ grains and the insulating strontium titanate layer (denoted by index 1), and the insulating intergranular layer (denoted by index 2); 1 — semiconducting SrTiO$_3$ (d_g, σ_0), 2 — insulating SrTiO$_3$ ($d_1/2$, ε_1, σ_1), 3 — the intergranular phase (d_2, ε_2, σ_2). Reprinted from Wernicke (1981b) by permission of the American Ceramic Society

creases because of the penetration of the additives into it ($\Delta m/m$). Simple calculations give

$$\frac{1}{\varepsilon_{\text{eff}}} = \frac{1}{\varepsilon_1}\frac{d_1}{d_g} + \frac{1}{3\varepsilon_2}\frac{\varrho_1}{\varrho_2}\frac{\Delta m}{m} \qquad (5.3)$$

where ϱ_1, ϱ_2 are the densities of the intergranular phase and the $SrTiO_3$ material, respectively.

The parameters of the diffusion $SrTiO_3$ layers (denoted by index 1) can be measured by measuring the properties of the sample after it is subjected to the oxidizing heat treatment, but before depositing oxide additives on its surface. Just as with PTCR resistors, we can use here the general equations given by Daniels et al. (1978/79). The value of ε_1 is taken to be the same as that valid for $SrTiO_3$. The parameters of the intergranular phase (denoted by index 2) are much more difficult to determine, since they depend on both the composition of this phase and the technique used for its formation. The value of the dielectric constant ε_2 lies between 50 and 300.

If the values of $\Delta m/m$ are sufficiently large, the first term in equation (5.3) may be neglected; if this is done, the variation of $1/\varepsilon_{\text{eff}}$ with $\Delta m/m$ becomes linear. Based on an experimental curve, Wernicke (1981b) calculated that $\varepsilon_2 = 200$ (assuming that $\varrho_1 = \varrho_2$).

The thickness d_2 of the intergranular phase decreases as the refiring temperature increases, since the viscosity of the phase that penetrates in-between the grains decreases. The effective dielectric constant then increases monotonically. At high temperatures (such as 1420 K), however, ε_{eff} begins to decrease again, since the diffusion layers are widened considerably. At 1420 K, the diffusion rate of the Sr vacancies also greatly increases.

Using the model described above we can control the manufacturing process so that the capacitors show the desired properties. In particular, we can gain much valuable information about the course of the process by measuring the variation of the sample masses (a measurement easy to perform). For example, if we wish to produce a device with a small temperature coefficient of resistivity, the relative effects of the intergranular and diffusion layers should strictly be controlled; we can achieve this by ensuring appropriate process conditions and by adding greater amounts of SiO_2 and Al_2O_3, whereby increasing the influence of the intergranular phase upon the properties of the capacitors. The proposed parameters are:
— starting material composition: $Sr_{100}(Ti_{100}Nb_{15})O_{300}$,
— additives: 0.1 wt% Al_2O_3, 0.2 wt% SiO_2,
— sintering carried out at 1720 K in H_2/N_2,
— refiring at 1370 K.

A disadvantageous effect appearing in this case was that the effective dielectric constant decreased to about 10 000. The properties of the material thus prepared were as follows: $\tan\delta = 6 \times 10^{-3}$, temperature/capacitance coefficient $\Delta C/C_{298\,K} = \pm 1\%$ (at 218–398 K), and $\varrho_{\text{eff}} = 10^{11}$ $\Omega \cdot$ cm at 1 kV/cm.

In another experiment, the objective was to obtain a material with a potentially large ε_{eff}, whereas the remaining parameters were to remain unchanged. The starting material was doped with a small amount of SiO_2 and the process parameters were:
— starting material composition: $Sr_{100}(Ti_{100}Nb_{10})O_{300}$,
— additives: 0.05 wt% SiO_2,
— sintering carried out at 1740 K in H_2/N_2,
— refiring at 1420 K.

The material obtained had the properties: $\varrho_{\text{eff}} = 65\,000$, $\tan\delta = 7 \times 10^{-3}$, $\Delta C/C_{298\,K} = \pm 22\%$ (at 218–398 K), $\varrho_{\text{eff}} = 10^{10}$ $\Omega \cdot$ cm at voltage of 1 kV/cm.

These experiments confirmed the validity of the proposed model. Waku (1967) examined the effect of the CuO acceptor additive upon the properties of GBBL capacitors. The resistivity of the material reached a maximum at the CuO concentrations at which the dielectric constant and $\tan\delta$ were minimum. The results obtained are shown in Fig. 5.6.

Waku found that the diffusion of Cu ions along the intergranular phase boundaries proceeds more quickly that conventional diffusion along the grain

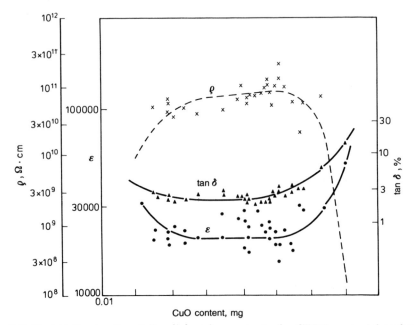

Fig. 5.6. Resistivity, $\tan\delta$ and the dielectric constant of a GBBL material as functions of the CuO acceptor content (after Waku, 1967)

boundaries. He observed that, on the surfaces of the $BaTiO_3$ grains, a thin layer of the solid solution of $BaTiO_3$ and CuO formed; the thickness of this layer was 1–2 μm and the average grain diameter was between 20 and 40 μm. The effect of CuO upon the properties of porous $BaTiO_3$ material was also examined by Kuwabara (1983).

Many other investigators published the results of their experiments in fabricating GBBL capacitors. New materials suitable for these capacitors are often proposed (Sakabe, 1987). Syamaprasad et al. (1987), for example, produced a material composed of $Ba_{0.80}Pb_{0.20}(Ti_{0.88}Zr_{0.12})$ doped with CaO (0.8 mol%) and ZrO_2 (1 mol%). The dielectric constant of this material was $\varepsilon = 3550$, and it varied by 5.7% within the temperature range from 283 to 333 K, $\tan\delta = 0.005$, and the resistivity was of the order of 1.3×10^{12} $\Omega\cdot$cm and varied linearly with temperature.

Igarashi et al. (1981) proposed using the gaseous phase formed when the sintered ceramic is fired together with powders of appropriate compositions ($PbZrO_3$, $Bi_2(ZrO_3)_3$, CuO, Al_2O_3). This gaseous phase was to be the source of the additives that contribute to the formation of the intergranular phase in $BaTiO_3$. This technology seems to be useful since it eliminates the impurities introduced when using the oxide additives.

Waser (1988) described the defect structure of barium titanate, paying special attention to the formation and solubility of the defects introduced by hydrogen. He has found that, in donor-doped materials, the effect of these defects upon the electrical properties is minor.

The technology of boundary layer capacitors provides a good example of how knowledge of the mechanisms of physical phenomena permits control of the structure of matter. This technology may form the basis for developing new technologies of other monolithic materials with various desired properties.

REFERENCES

Brauer H. and Kuschke R. (1971), Dielectric capacitors with inner barrier layers and low temperature dependence, US Patent 3,569,802, March 9.

Brauer H. (1974), Resistance anomaly in semiconductor $BaTiO_3$ ceramics in range below Curie point, *Solid State Electron.*, **17** (10), 1013.

Daniels J., Hardtl K. H. and Wernicke R. (1978/79), The PTC effect of barium titanate, *Philips Tech. Rev.*, **38** (3), 73.

Glaister R. M. (1962), Dielectric ceramic compositions and the method of production thereof, US Patent 3,028,248, April 3.

Goodman G. (1981), Capacitors based on ceramic grain boundary barrier layers. A review, in: Levinson L. M. and Hill D. C. (eds.) (1981), *Grain Boundary Phenomena in Electronic Ceramics. Advances in Ceramics*, Vol. 1, Am. Ceram. Soc., Columbus, Ohio, 215.

Igarashi H., Betoh C. and Okazaki K. (1981), Vapor phase diffusion of metal oxides into $BaTiO_3$ ceramics and its application to multilayer boundary layer capacitors, in: Levinson L. M. and Hill D. C. (eds.) (1981), *Grain Boundary Phenomena in Electronic Ceramics. Advances in Ceramics*, Vol. 1, Am. Ceram. Soc., Columbus, Ohio, 254.

Klerk J. and Sanders P. J. H. (1981), $SrTiO_3$ boundary layer capacitors: Influence of additives and stoichiometry, in: Levinson L. M. and Hill D. C. (eds.) (1981), *Grain Boundary Phenomena in Electronic Ceramics. Advances in Ceramics*, Vol. 1, Am. Ceram. Soc., Columbus, Ohio, 282.

Kuwabara M. (1983), Effect of CuO addition on the electrical an dielectrical properties of semiconducting barium titanate ceramics, *J. Mat. Sci. Lett.*, **2** (8), 403.

Philips Data Handbook (1984), Components and Materials, Part 11.

Roup R. R. and Butler C. E. (1950), Layerized high dielectric constant piece for capacitors and process of making, US Patent 2,520,376, Aug. 29.

Sakabe Y. (1987), Dielectric materials for base-metal multilayer ceramic capacitors, *Am. Ceram. Soc. Bull.*, **66** (9), 1338.

Stenton N. and Harmer M. P. (1983), Electron microscopy studies of a strontium titanate based boundary layer material, in: Yan M. F. and Heuer A. H. (eds.) (1983), *Additives and Interfaces in Electronic Ceramics. Advances in Ceramics*, Vol. 7, Am. Ceram. Soc., Columbus, Ohio, 156.

Sundaram S. K. (1990), Complex-plane impedance analysis of PTC thermistor-intergranular capacitor transition, *J. Phys. D, Appl. Phys.*, **23** (1), 103.

Syamaprasad U., Galgali R. K. and Mohanty B. C. (1987), A modified barium titanate for capacitors, *J. Am. Ceram. Soc.*, **70** (7), C-147.

Waku S. (1967), Studies on boundary layer ceramic capacitor (BL capacitor most suitable high frequency by-pass capacitor), *Rev. Electric. Commun. Lab.*, **15** (9–10), 689.

Waku S. (1970), Studies on boundary layer ceramic capacitor, *J. Phys. Soc. Jpn.*, **28** (suppl.), 457.

Waku S., Uchidate M. and Kiuchi K. (1970), Studies on (Ba, Sr)TiO_3 boundary layer ceramic dielectrics (Development of BL capacitor for submarine coaxial cable repeater), *Rev. Electric. Commun. Lab.*, **18** (9–10), 681.

Waku S., Nishimura A., Murakami T., Yamaji A., Edahiro T. and Uchidate M. (1971), Classification and dielectric characteristics of boundary layer ceramic dielectrics (BL dielectrics), *Rev. Electric. Commun. Lab.*, **19** (5–6), 665.

Waser R. (1988), Solubility of hydrogen defects in doped and undoped $BaTiO_3$, *J. Am. Ceram. Soc.*, **71** (1), 58.

Wernicke R. (1981a), Formation of second-phase layers in $SrTiO_3$ boundary layer capacitors, in: Levinson L. M. and Hill D. C. (eds.) (1981), *Grain Boundary Phenomena in Electronic Ceramics. Advances in Ceramics*, Vol. 1, Am. Ceram. Soc., Columbus, Ohio, 261

Wernicke R. (1981a), Two-layer model explaining the properties of $SrTiO_3$ boundary layer capacitors, in: Levinson L. M. and Hill D. C. (eds.) (1981), *Grain Boundary Phenomena in Electronic Ceramics. Advances in Ceramics*, Vol. 1, Am. Ceram. Soc., Columbus, Ohio, 272.

Yamaoka N. and Matsui T. (1981), Properties of $SrTiO_3$-based boundary layer capacitors, in: Levinson L. M. and Hill D. C. (eds.) (1981), *Grain Boundary Phenomena in Electronic Ceramics. Advances in Ceramics*, Vol. 1, Am. Ceram. Soc., Columbus, Ohio, 232.

Yamaoka N., Masuyama M. and Fukuji M. (1983), $SrTiO_3$-based boundary layer capacitor having varistor characteristics, *Am. Ceram. Soc. Bull.*, **62** (6), 699.

Yan M. F. and Heuer A. H. (eds.) (1983), *Additives and Interfaces in Electronic Ceramics. Advances in Ceramics*, Vol. 7, Am. Ceram. Soc., Columbus, Ohio.

CHAPTER 6

Ceramic Gas Sensors

The general definition of a sensor says that it is a device which, in response to a change of a selected external factor, changes at least one of its parameters. This definition may, however, be used for describing most ceramic components employed in electronics. These external factors may, for example, include temperature, electric voltage, the presence or a given concentration of a gas in an atmosphere, and so on. The sensors reacting to the first two factors, i.e., varistors, capacitors and thermistors, have been discussed in previous chapters. Those reacting to the presence or concentration of a gas will be called gas sensors. The characteristic property of the gas sensors that is most often measured is their electrical conductivity. It is a very convenient parameter, since its value is unequivocal and easy to measure. In the present book we shall not deal with the electronic circuits in which the gas sensors may be operated. Reviews and classification of gas sensors have been given by Bühling (1985), Ichinose (1985, 1986, 1987), Kulwicki (1984), Moseley (1992), Saaman and Bergveld (1985), and Uchino (1986).

Gas sensors, including ceramic devices, are applied in many fields of science and engineering, such as:

— measurements of the content of inflammable or poisonous gases in the air or other atmospheres,

— alarm installation in mines and factories,

— examination of exhaust gases, measuring the contents of harmful gases, such as CO and nitrogen oxides,

— measurements of the fuel/air ratio in engines with the aim of controlling the fuel mixture,

— home appliances, such as sensors installed in microwave ovens, or fire sensors.

Among the examples of recent applications of ceramic gas sensors we can mention the monitoring of combustion processes and CO detection (Torvela et al., 1989, 1990), fire detection in wooden houses (Amamoto et al., 1990), and the detection of gaseous carbon tetrachloride (Torvela et al. 1991b).

Ichinose (1985, 1986, 1987) has proposed classifying semiconducting ceramic components into the following groups:

(1) components utilizing the properties of the grain bulk, such as the NTCR (negative temperature coefficient of resistivity) thermistors, high-temperature thermistors, oxygen ZrO_2-based sensors;

(2) components utilizing the properties of the grain boundaries, such as the PTCR (positive temperature coefficient of resistivity) thermistors, boundary-layer capacitors, ZnO-based varistors;

(3) components utilizing the properties of the surface of a ceramic material, such as gas sensors, humidity sensors, catalysts.

The gas sensors belonging to the third group are fabricated of n-type semiconducting ceramic materials, such as SnO_2, ZnO, TiO_2, and Fe_2O_3. They are manufactured using a variety of techniques, among which the three most widely employed are:

— thin-layer techniques (vacuum evaporation, sputtering and CVD methods),

— thick-layer techniques,

— sintering techniques.

In the present book we shall chiefly deal with the last two groups, although occasionally (when discussing the theory of operation of the sensors) we resort to the results obtained for the first group. Most attention will be devoted to SnO_2-based sensors, since they are most widely used.

The general principle of operation of gas sensors is as follows. The ceramic material, after sintering it to a low density, should be fine-grained ($\bar{d} < 1$ µm), so that the surface area of contact with the gases to be controlled is the greatest. When the sensor is placed in clean air, the oxygen adsorbed by it increases its resistance (the resistance chiefly depends on the surface condition of the sensor, since its individual grains only touch one another through the narrow bridging necks). As the concentration of the reducing gas (e.g., CO, methane, propane or higher hydrocarbons) contained in the atmosphere increases, it reacts with the adsorbed oxygen, thereby decreasing the resistance of the surface and, thus, of the whole device.

Gas sensors must fulfill many exploitation requirements: the most important parameters are:

— sensitivity (high enough but below a certain level so as to avoid false alarms);

— selectivity (e.g., an alcohol sensor must not react to methane);

— reading reproducibility (each unit should indicate the same value, and this should be unchanged after subsequent switch on/off cycles);
— reading stability during operation (sensors are often operated in an intermittent way for several years);
— quick response (typically, less than 10 s);
— small size;
— safety of operation, especially when operated in the presence of inflammable or explosive gases;
— low power consumption (usually between 0.5 to 1.0 W);
— preferably a linear conductivity variation, so that the necessary electronic circuits can be simple;
— low costs.

The parameters that are most important and most difficult to achieve are the sensitivity and selectivity of the sensors. Until now few manufacturers in the world have satisfied these requirements.

We shall now discuss the achievements in manufacturing and examining the properties of gas sensors. First we shall describe their technology and basic characteristics and, then, we shall present the varying opinions of various investigators about the operating mechanism of the sensors.

6.1 TECHNOLOGY, DESIGN AND PROPERTIES OF SnO_2-BASED SENSORS

As already mentioned, only a few manufacturers have succeeded in fabricating good-performance gas sensors. One of these few is the Japanese company Figaro Inc. Watson and Yates (1985) tell the story how the Figaro Inc. began to work on the technology of gas sensors. In 1961, a strong gas explosion took place in a factory situated near the Kawaguchi lake in Japan, killing eight employees. This tragic event motivated engineers to think how to design a reliable and cheap gas sensor. The first such sensor was designed and constructed (of a ZnO-based material), during one day only, by Naoyoshi Taguchi, a mechanical engineer, who later founded Figaro Inc. in 1962. Over a few years, the company developed the SnO_2-based 109 TGS (Taguchi gas sensors) type (Taguchi, 1972). In 1970, the production volume of these sensors amounted to 700 000 devices per year. Since 1975, the Figaro Inc. have been manufacturing 1.8 million per year of new 812-series sensors (sensitive to CO, isobutane, hydrogen and ethanol) and 813-series sensors reacting to methane. The basic configuration of the Figaro 813 gas sensor is shown in Fig. 6.1.

The carrier element of the sensor consists of a thin-walled ceramic tube with an internal diameter of the order of 1 mm. It is made of Al_2O_3 ceramic

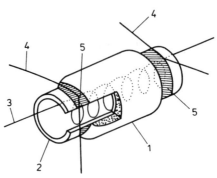

Fig. 6.1. Configuration of the Figaro gas sensor; 1 — sintered SnO_2, 2 — ceramic tube, 3 — heater filament, 4 — lead, 5 — electrode

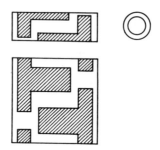

Fig. 6.2. Figaro gas sensor; configuration of the electrodes deposited on the ceramic tube

of high thermal conductivity (of the order of 20 W/(m·K)). The electrodes (Au, Pt) are deposited upon the tube in a way shown in Fig. 6.2. In order to ensure the stability of the sensor readings, the electrodes should show a high resistance to the action of ambient atmosphere. The electrode metallization may, for example, be effected by firing an appropriate metallic (Au, Pt) paste. The tube prepared in this way is covered with a ceramic paste (doped SnO_2), whose resistance varies as a function of the temperature and composition of ambient atmosphere. The tin oxide powder is obtained from $SnCl_4$, which is first converted into stannic acid by treating it with an aqueous solution of ammonium hydroxide and, then, calcined. The paste is also doped with catalysts, which are most often Pt or Pd powders introduced in the form of aqueous $PdCl_2$. Sometimes the paste is also doped with Al_2O_3. This paste is sintered at a temperature between 970 and 1270 K in air. The exact chemical composition and the detailed technology of the commercial gas sensors manufactured by individual manufacturers are confidential. After the paste is sintered, it may be impregnated with an appropriate substance. Inside the tube, a platinum heater filament (30 Ω) is installed. The electrodes are bonded to the leads which, together with the heater endings, are welded to an appropriate pin (made of, e.g., Ni or Kovar) of a multipin housing. The housing has usually a standard pin arrangement, compatible with electronic components sockets.

The plastic materials of which the sensor housings are made must be resistant to the aggressive gaseous environments and to high temperatures (of the order of 470–620 K), conditions under which the sensors often operate. One such material is Nylon-66. The atmosphere to be controlled enters the sensor through a window covered with a double layer of 100 mesh stainless steel gauze. This gauze prevents the inflammable gases from igniting when getting in contact with the active element operated at an elevated temperature. Watson and Yates (1985) report that even a spark produced inside the sensor housing protected with such a gauze does not ignite a hydrogen-oxygen mixture (2:1) which is known to be very explosive.

Some sensors must first be activated and stabilized before they are installed for normal operation. For example, oxygen-sensitive ZnO-based sensors are activated by heating them in hydrogen. The behaviour of the Figaro sensor in the initial stage of operation is shown in Fig. 6.3. As we can see,

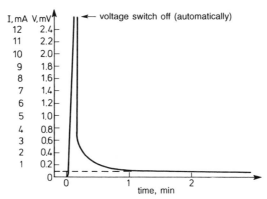

Fig. 6.3. Transient overvoltage occurring in the initial stage of the operation of the Figaro gas sensor (after Watson, 1984; reproduced by permission of Elsevier Sequoia S.A.)

it reaches full stabilization after as little as 1 min. When a sensor is heated up to its operating temperature (about 620 K), its conductivity sharply increases because of the removal of the substances adsorbed during storage. Obviously, the time needed for stabilization depends on the conditions and time of storage.

After this initial period, the readings of the sensor are stable. For example, the resistance of 812 sensor (which is of the order of several kΩ) placed in an air atmosphere (at 293 K and 65% relative humidity) that contains 1000 ppm isobutane, decreases by about 0.5 kΩ during the initial 2 days of operation and, then, remains constant for the next 12 days (Watson, 1984). It is recommended that a sensor should be stabilized for a week before it is operated. It should also be recalibrated at appropriate time intervals. The calibration should be carried out under strictly specified conditions, in particular, temperature and humidity. Figure 6.4 shows the typical variations of the conductance of the Figaro 812 gas sensors versus temperature and humidity (Watson, 1984). The effect of temperature upon the sensor conductance during operation may be compensated using an appropriate

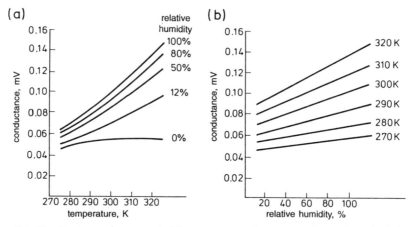

Fig. 6.4. Conductance (represented by an appropriate measuring voltage) of the Figaro gas sensor (812, 813, 711-types) as a function of temperature (a) and humidity (b); reprinted from Watson (1984) by permission of Elsevier Sequoia S.A.)

electronic thermistor-based circuit. The compensation may also be effected using a humidity sensor and then re-calculating the readings of the basic sensor. The Figaro 815D gas sensor, recently introduced on the market, is almost humidity-insensitive.

Another factor that may cause spurious readings of the sensor is the rate at which the air flows through it. An effective remedy here is the flame-protecting steel gauze.

Figures 6.5 and 6.6 show typical variation of the conductance of various Figaro sensors as a function of CO and CH_4 concentrations. The 711 version has been designed especially for detecting CO. We can see that the selectivity of all the sensors is high.

Clifford and Tuma (1982/83a,b) describe many parameters of the Figaro sensors. They measured these parameters at various concentrations of oxygen, nitrogen, hydrogen, methane, carbon oxide, steam and their mixtures. They also examined how these parameters varied with the measurement temperature.

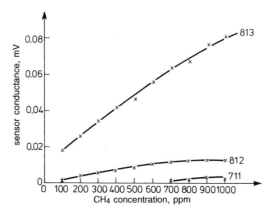

Fig. 6.5. Conductance (represented by an appropriate measuring voltage) of the Figaro 812, 813, 711 gas sensors as a function of CH_4 concentration (reprinted from Watson, 1984, by permission of Elsevier Sequoia S.A.)

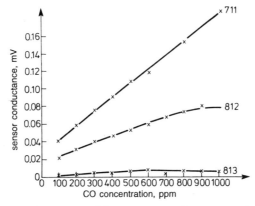

Fig. 6.6. Conductance (represented by an appropriate measuring voltage) of the Figaro 812, 813, 711 gas sensors as a function of CO concentration (reprinted from Watson, 1984, by permission of Elsevier Sequoia S.A.)

Many authors describe the properties of ceramic gas sensors fabricated by them using different techniques. Below, we discuss some of this work.

The technique of preparing and depositing the tin oxide paste used by Oyabu et al. (1986) is shown in Fig. 6.7. The starting SnO_2 powder with a purity of 99.99% and a grain diameter of 10–100 μm was milled until the grain diameter was reduced to 1–10 μm. Then, after drying at a temperature of 390 K for 2–3 hours, a catalyst (Pt or Pd) was introduced into it by stirring it with distilled water. The paste was deposited onto a ceramic tube (1.2 mm diameter, 5 mm long), fired at 570 K and then sintered in air at a temperature of 1070 K. In the next step, the gas-sensing layer of the element was impregnated with silica sol $(H_n Si(OH)_{4-n})$ and sintered for up to 30 min

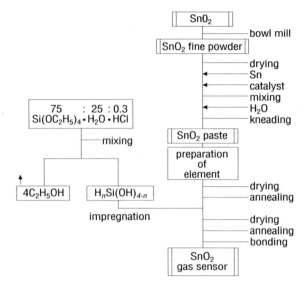

Fig. 6.7. Flow chart for preparation of sintered tin oxide sensor (reprinted from Oyabu et al., 1986, by permisssion of Elsevier Sequoia S.A.)

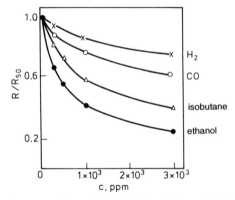

Fig. 6.8. Resistance of sensors (doped with 0.2% platinum black) fabricated by Oyabu et al. (1986) as a function of the concentration of various gases at a temperature of 520 K (reprinted by permission of Elsevier Sequoia S.A.)

at 1020 K. The parameters of the sensors thus obtained were measured after ageing at 620 K for 15 days. The sensors were sensitive (their resistance changed) to all gases tested: ethanol, CO, isobutane and H_2, as shown in Fig. 6.8. The symbol R_{50} marked in the figure denotes the resistance of the sensor when the concentration of the gas to be detected in a given atmosphere was 50 ppm. The selectivity of these sensors was not very high.

It will be shown later that noble metal catalysts greatly improve the sensitivity and selectivity of ceramic gas sensors. The activity of the catalysts strongly depends on the size of their particles and their dispersion in the ceramic material. Matsushima et al. (1992) examined various methods of incorporating Pd in an SnO_2 sensor, such as the standard impregnation with an acid solution of $PdCl_2$, the deposition from a chloro-complex and the colloid method. In the second method (also known as the fixation method), an SnO_2 powder was suspended in a solution that contained a $[PdCl_4]^{2-}$ complex (a solution of $PdCl_2$ and NH_4Cl in water). After the complex reacted with the surface hydroxyl, the powder (after filtration) was annealed at 570 K in a hydrogen flux to convert the surface deposits into metallic Pd. In the colloid method, an SnO_2 powder was suspended in a diluted aqueous colloid of Pd from which the Pd particles were deposited on the powder surface. Transmission electron microscopy observations shown that the fixation method gives the finest dispersion of palladium with a mean particle size of 1.6 nm, and the sensors prepared using this method had a better sensitivity than those produced by the two other methods.

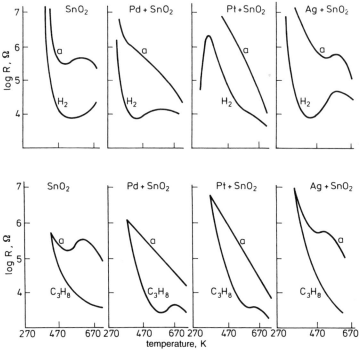

Fig. 6.9. The behaviour of a gas sensor in various atmospheres depending on the kind of metallic catalyst added to the material; the catalyst content 0.5 wt%; gas concentrations: H_2 — 0.8% by vol., C_3H_8 — 0.2% by vol.; a — air (reprinted from Yamazoe et al., 1983, by permission of Elsevier Sequoia S.A.)

Yamazoe et al. (1983) fabricated SnO_2-based sensors of various designs. A metallic tin powder (99.999%) was washed with hot nitric acid. The precipitate thus produced was washed with water, dried at 370 K and, then, calcined in air at 870 K for 10 h. The metallic catalyst Pd, Pt or Ag were added (0.5 wt%) in the form of chlorides or nitrates. A pellet 10 mm in diameter and 1 mm thick was pressed under a pressure of 150 MPa, and two 0.3 mm Pt wire electrodes were pressed into it at a distance of 4 mm from one another. The pellet was fired in air at 870 K for 5 h. The resistance of the sensors obtained in this way, but doped with different substances, was measured as a function of temperature in various air atmospheres that contained 0.8% H_2 or 0.2% C_3H_8. The results obtained are shown in Fig. 6.9. By analysing the shapes of the curves shown in this figure, we can describe the mechanism of the catalysis that occurs during the operation of the sensor. We shall discuss this in Section 6.3.

Coles et al. (1985) fabricated gas sensors using a mixture of SnO_2, 36 wt%

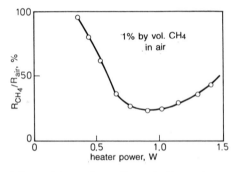

Fig. 6.10. The ratio of the resistance in methane to the resistance in air, R_{CH_4}/R_{air}, of a ceramic gas sensor as a function of the heater power (reprinted from Coles et al., 1985, by permission of Elsevier Sequoia S.A.)

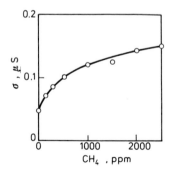

Fig. 6.11. Conductance of the gas sensor doped with aluminium silicate, designed by Coles et al. (1985), as a function of methane content (reproduced by permission of Elsevier Sequoia S.A.)

aluminium silicate and 1.55% $PdCl_2$. Mixing it with distilled water, they obtained a dense paste, which was then painted on a sintered aluminium oxide plate with two Pt electrodes deposited on it. On the other side of the plate, a Pt paste heater was painted. After drying, the sensor was fired at 1270 K for 2 h. In the same way, these investigators fabricated sensors intended for detecting carbon monoxide, except that the aluminium silicate was replaced by 15% Bi_2O_3, and the firing temperature was reduced to 1070 K. The temperature of the sensor in operation was difficult to measure, since it is proportional to the heater power. Figure 6.10 shows how the ratio of the sensor resistance in methane (1% by volume) to its resistance in air, R_{CH_4}/R_{air}, varies with the heater power. The optimum heater power was chosen to be 0.9 W. The heater temperature was then about 470 K. The conductance of the sensor varied with the methane concentration as shown in Fig. 6.11; the sensitivity appeared to be highest at low concentrations of methane. When Bi_2O_3 was added, the sensitivity of the sensors to carbon monoxide increased considerably. Just as in the preceding experiment, the optimum heater temperature was determined; the heater power appeared to be 0.9 W again. Figure 6.12 shows the variation of the sensor conductance as a function of the CO and CH_4 concentrations. We can see from the figure that the sensor shows a high selectivity, and is able to distinguish between the presence of carbon monoxide and methane. The sensor even reacted (by changing its conductance) to the presence of 12.5% by vol. of CO (which corresponds to the explosive concentration of this gas in air).

As the Bi_2O_3 concentration in the material increased, its resistance in the

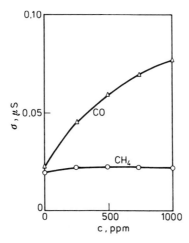

Fig. 6.12. Conductance of the sensor doped with bismuth oxide, designed by Coles et al. (1985), as a function of methane content (reproduced by permission of Elsevier Sequoia S.A.)

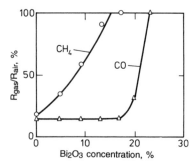

Fig. 6.13. Selectivity of SnO_2-based sensor as a function of the bismuth oxide additive content (reprinted from Coles et al., 1985, by permission of Elsevier Sequoia S.A.)

CH_4 (1% by vol.) atmosphere greatly increased, whereas in the CO (1% by vol.) atmosphere it remained unchanged until Bi_2O_3 concentration reached about 17 wt% (see Fig. 6.13). The selectivity was, thus, highest when the Bi_2O_3 concentration was between 15 and 16 wt%.

Table 6.1 gives the resistance values of these sensors measured in other gaseous atmospheres (1% by vol. of the gas to be detected). These values suggest that:

Table 6.1 Resistance responses (in MΩ) of SnO_2 sensors: CH_4-sensitive sensor (doped with aluminium silicate) and CO-sensitive sensor (doped with Bi_2O_3) to various gases (1% by vol. in clean air) (after Coles et al., 1985)

Gas	CH_4-sensitive sensor	CO-sensitive sensor
Clean air	19.5	47.0
Carbon monoxide	18.9	9.6
Methane	3.4	46.2
Ethane	1.8	36.4
Propane	1.2	28.6
Butane	0.8	23.3
CCl_2F_2	17.2	32.7
$CHClF_2$	7.5	5.2**
Hydrogen	0.7	4.6
Acetic acid vapour*	0.2	0.8
Nitric acid vapour*	5.0	8.3
Sulphuric acid vapour*	14.0	11.8
Hydrochloric acid vapour*	0.3	3.0
Ammonia vapour*	0.2	0.2
Chlorine gas*	40.0	> 100**
Dimethyl silane*	0.002	0.002**

* Approximate concentrations.

** Resistance change irreversible except by high power heat cleaning or exposure to clean air for several tens of minutes.

— with increased length of the carbon chain, the sensitivity of the sensor increases, probably because of the increasing number of hydrogen atoms ready to take part in the reaction;

— it is possible to measure, in a reversible manner, the concentration of acids, silanes and NH_3;

— as a result of the adsorption of chlorine, the conduction electrons become bound in a permanent manner, which probably happens through the dissociation of the Cl_2 molecule into the Cl^- ions.

A further improvement in sensor properties has recently been achieved by doping with Sb_2O_3 (a reduction in the sensor resistance), and by modifying the heat-treatment procedure (Coles et al., 1991a,b). Doping with Sb_2O_3 was also studied by Paria and Maiti (1982) and Uematsu et al. (1987).

Romppainen et al. (1985) examined how the presence of CH_4, SO_2 and NO affect the behaviour of the SnO_2-based sensors designed to detect CO. These examinations were aimed at simulating the conditions that prevail in the exhaust gases of combustion engines. In the experiment, the SnO_2 paste was added with glass in order to increase its mechanical strength after sintering. Ethyl cellulose and terpineol were used as the dispersant medium. The paste was deposited sequentially on an Al_2O_3 substrate using the screen-printing technique until it reached a thickness of 30 µm (after firing). Prior to the screen-printing operation, two Au electrodes spaced at 0.5 mm from one another and fitted with Pt leads were deposited on the Al_2O_3 substrate. The whole structure was fired at 1220 K. Its parameters were measured in a special chamber that contained an appropriate atmosphere at a controlled temperature. The atmosphere was composed of 80 mol% N_2, 15 mol% CO_2 and 5 mol% O_2 added with controlled amounts of CO, CH_4, SO_2 or NO (in order to simulate the composition of combustion gases free of steam). The measurements were made at 770 K, so as to ensure the quick recovery of the sensor. Figure 6.14 shows the variation of the electric current that flows through the sensor as a function of the concentration of the various gases contained in the synthetic combustion gas. The sensor appeared to be most sensitive to CO. Figure 6.15 shows some variations of the CO-detection sensitivity due to the presence of other gases. The sensor seems to be useful for detecting the CO concentration in combustion gases.

Yasunaga et al. (1986) report that ethyl orthosilicate added to the sensor material improves (according to its degree of polymerization) the long-term sensitivity and stability of the sensor. Figure 6.16 lists the procedures involved in the preparation of these sensors. The authors do not specify the parameters of the heat treatment to which the sensors are subjected. Polymerized ethyl orthosilicate used as the binder was obtained from the mixture composed of the monomers of this compound with ethylene and HCl in var-

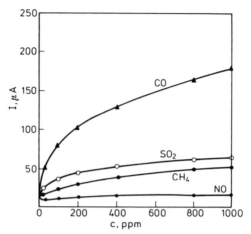

Fig. 6.14. Electric current flowing through the Romppainen gas sensor as a function of the concentration of various gases in the synthetic combustion gas. The current flowing in a pure combustion gas is 13 µA (reprinted from Romppainen et al., 1985, by permission of Elsevier Sequoia S.A.)

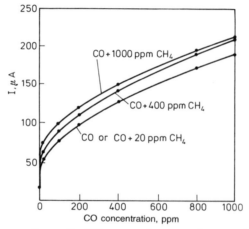

Fig. 6.15. Spurious readings of a CO sensor due to an increased CH_4 concentration (reprinted from Romppainen et al., 1985, by permission of Elsevier Sequoia S.A.)

ious proportions, so as to obtain various values of DP (i.e., the degrees of polymerization) determined from viscosity measurements. Inside the sensor former, two Pt–Ir spirals were installed to function as the electrodes, with one of them also acting as the heater.

Figure 6.17 shows how the resistance of the sensor varies with temperature in various atmospheres, at two different polymerization degrees of the binder. As the polymerization degree increases, the sensitivity of the sensor increases and its resistance decreases. The initial time required for the sen-

Sec. 6.1] SnO$_2$-based Sensors 173

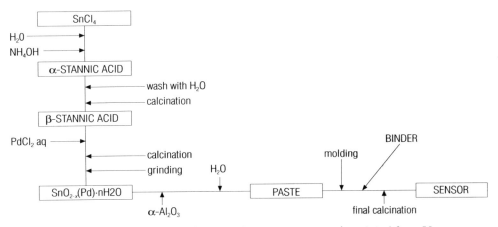

Fig. 6.16. Preparation of an SnO$_2$ ceramic-type gas sensor (reprinted from Yasunaga et al., 1986, by permission of Elsevier Sequoia S.A.)

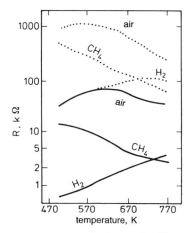

Fig. 6.17. Resistance of the gas sensor designed by Yasunaga et al. (1986) versus temperature in various atmospheres and at various polymerization degrees (DP) of the binder (dotted line DP = 2.5, solid line DP = 3.4); the gas concentration was 0.2% by vol. (reproduced by permission of Elsevier Sequoia S.A.)

sor to stabilize also increases. At DP < 2.5, the long-term stabilization of the sensor increases.

A sensor designed to operate in an alarm system requires an appropriate data analysing and processing electronic circuit. The resistance of the sensor must be carefully matched to its predicted resistance changes in the given system to be monitored. If the sensor resistance is, for example, too great (of the order of megaohms), the system may respond to unwanted external factors. The sensor resistance may be adjusted most easily by appropriate doping of the semiconducting starting material. We have already mentioned such dopants as aluminium silicate, Bi_2O_3 (which increase the sensor resistance), or donor-type dopants, such as Sb_2O_3. Some investigators suggest that these additives also affect the selectivity, stability, and other properties of the sensor.

Yannopoulos (1987) examined how an addition of antimony oxide affects the properties of thick-layer SnO_2 sensors. The preparation of doped $Sn_{0.98}Sb_{0.02}O_2$ by annealing an appropriate powder mixture, then milling it and refiring until it becomes homogeneous is time- and energy-consuming, and the composition of the compound may change since its individual components differ in their vapour pressures. Moreover, it is difficult to introduce the required small amount of the dopant during a conventional ceramic process. For these reasons, Yannopoulos prepared the material by precipitating a mixture of Sn and Sb hydroxides added in the form of liquid $SnCl_4$ and $SbCl_5$ to ammonium hydroxide. The precipitates thus obtained were then centrifuged, dried and calcined twice at a temperature of 1070 K in air (the second calcination was carried out for 51.5 h). The paste was doped with 5 wt% Pd (by adding $PdCl_2$) and 5 wt% magnesium oxide, and, in one experiment, also with 9 wt% nickel silicide, resistant to oxidation. Then the paste was deposited on a forsterite substrate fitted with two Pt wire electrodes.

The properties of the sensors were measured using a resistance furnace through which an appropriate gas was passed. The sensors were first annealed in dry air at a temperature of 920 K. Then, the voltage drop across the sensor was measured (at a constant current) at various concentrations of H_2 and CO in the reference gas whose composition was: 2% O_2+24% H_2O+N_2 (it simulated the concentration of the gas furnace gases). Figure 6.18 shows the time-variation of this voltage at various H_2 and CO concentrations. After the concentrations of these gases were changed, the voltage across the sensor changed by 90% of its total change during the first 30 s of the operation of the sensor. The resistance of the sensor at the measurement temperature (770 K) was a few ohms. The observed voltage variations were fully reversible, as the H_2 and CO concentrations increased and decreased. This reversibility of the sensor behaviour permitted continuous stable monitoring of the composition of the furnace gases.

McAleer et al. (1985, 1986) describe another way in which the SnO_2-based sensor may be used for measuring the gas concentration. They utilized the Seebeck effect, in which the thermoelectric force E is induced at the contact of two materials kept at different temperatures. This thermoelectric force is defined as

$$E = \frac{dV}{dx} = \alpha \frac{dT}{dx} \tag{6.1}$$

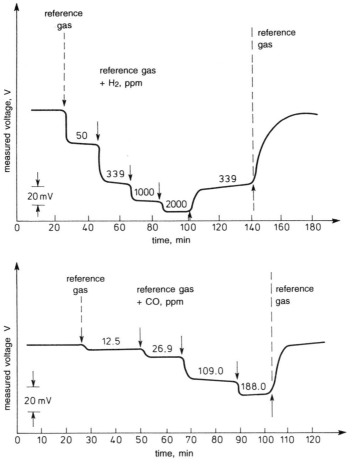

Fig. 6.18. Typical values of the output measuring voltage for the Yannopoulos (1987) sensor; the voltage varies with time as a function of the varying concentration of hydrogen and carbon monoxide present in the reference gas maintained at 770 K (reproduced by permission of Elsevier Sequoia S.A.)

where α is the Seebeck coefficient. McAleer et al. prepared a junction of two pellets, one made of SnO_2 and the other of SnO_2 (Pt, Pd). When the air containing 1% of hydrogen (at room temperature) was passed through the device, the temperature in the latter material increased by about 4 K. The Seebeck voltage then established was about 0.5 mV and decreased with decreasing hydrogen concentration. The increase in the temperature is attributed to the catalytic combustion of the hydrogen. These sensors show a low power consumption and the necessary electronic circuits are relatively simple.

In view of the various configurations and technologies of gas sensors, described above, we can see that research work in this field is continued. Mass production of ceramic gas sensors (now manufactured not only by the Figaro Inc.) has already been developed. Yasunaga et al. (1986) report that, in 1986, 12 million Japanese families used gas detectors for housekeeping purposes. The properties of the sensors have constantly been improved, based on the increasing understanding of their operating mechanism.

6.2 OPERATING MECHANISM

In Section 2.2, we have described the properties and the energy structure of the surface of the n-type semiconductor. Both ZnO and SnO_2 are such non-stoichiometric (due to the presence of excessive metal ions) semiconductors. As a result of surface defects and the adsorption of gases, a double layer forms on the surface: the depletion or accumulation layer.

After the fabrication process of a sensor is completed, its grain surfaces adsorb oxygen ions. This effect is described in more detail in Section 2.4. One of the possible reactions that then occur is

$$O_2 + 2e^- \to 2O^- \tag{6.2}$$

As a result, the free grain surfaces are covered with atomic layers of oxygen ions. The charge transfer occurring during this reaction leads to the formation of a potential barrier at the surface of the n-type semiconductor. If the ambient atmosphere contains other gases, such as RH_2 (in general notation), a dynamic equilibrium is established between reaction (6.2) and the reaction

$$RH_2 + 2O^- \to RO + H_2O + 2e^- \tag{6.3}$$

Since the height of the potential barrier is determined by the amount of transferred charge, it follows from equations (6.2) and (6.3) that the concentration of ions O^- on the surface determines the sensor conductance (when the ceramic is weakly sintered, the conduction chiefly occurs through the 'necks' between its grains). If the atmosphere does not contain RH_2 reducing substances, the number of adsorbed O^- ions is greater and the sensor shows a greater resistance (Morrison, 1987). As the concentration of RH_2 increases, the sensor resistance decreases. The energy structure of the crystal surface thus chiefly depends upon the composition of the atmosphere, since it is determined by the kind of the adsorbed ions (Fig. 6.19) and by temperature. The microstructure of a sintered ceramic material (not containing any catalyst) and its energy structure are shown in Fig. 6.20, where we can see the grains of a powder which has been very weakly sintered; it has been assumed that each grain contains a great number of electrons,

Sec. 6.2] Operating Mechanism

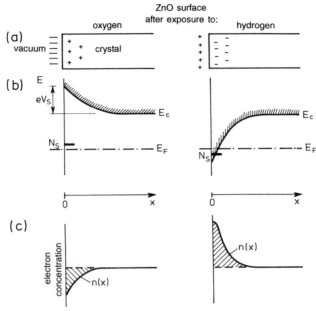

Fig. 6.19. The potential barrier (depletion layer) and the accumulation layer formed due to oxygen or hydrogen acting upon the surface of a ZnO crystal: (a) electric charge regions, (b) energy band model, (c) charge distribution; N_S — surface states

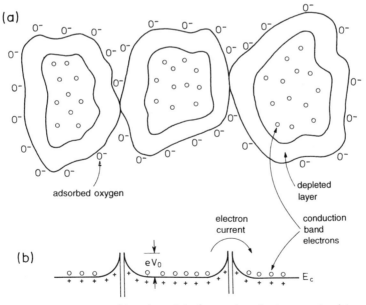

Fig. 6.20. Microstructure and band model of a semiconductor powder (the potential barriers form as a result of oxygen adsorption): (a) physical model, (b) energy band model (reprinted from Morrison, 1987, by permission of Elsevier Sequoia S.A.)

sufficient to fill all the surface states. Each time when moving from one grain to another, the electrons must overcome the potential barrier that exists at the grain boundary (cf. Section 2.3). But these electrons only form one component of the flowing current. The other component is the current that flows over the surface of the grains.

The surface conductance of the material is a function of the number and mobility of the carriers. The total conductance of the specimen as a whole is determined by both its surface and bulk conductances, which are connected in parallel one to another. When considering the sensor as a whole, we must also take into account the properties of the contacts between the metal electrodes and the semiconductor.

Heiland (1982) observes that in thin fine-grained layers the grains may not contain a sufficiently great number of electrons to build a complete space charge layer. If this is so, the whole thin layer contains space charge. A similar situation may occur in a porous ceramic or in a thick ceramic layer if their grains are small. In highly porous materials, the gas may easily penetrate into the material bulk, so that each grain should be considered individually.

Both the bulk and surface conductances may be adjusted according to the requirements. The bulk conductance depends on the dopants that form trap levels, whereas the surface conductance is affected by the energy structure, by the composition of the material that determine its ability to adsorb gases, and by the presence of catalyst that enhance this adsorption. The catalyst usually consists of small grains of metallic platinum or palladium. Adsorption and catalysis were discussed in detail in Section 2.4. Physical adsorption proceeding without charge transfer between the adsorbent and the crystal only slightly affects the conductance of the surface layer. It does, however, reduce the active surface area of the crystal and, thus, its chemisorption ability. The conductance of the sensor thus depends on the presence of large molecules, such as hydrocarbons.

To break some molecules so as to produce atoms that take part in chemisorption, energy must be delivered. The simplest way in which this may be done, especially when catalysts are present, is to increase the temperature. An example may be the reaction of the surface of zinc oxide to the presence of hydrogen (Heiland, 1957). At room temperature, hydrogen (0.1 MPa) does not affect the ZnO conduction. When, however, a number of H_2 molecules is decomposed at the hot tungsten filament, the ZnO surface conductance sharply increases due to chemisorption. This effect is utilized for detecting hydrogen present in vacuum, and atomic hydrogen present in a H_2 atmosphere (Haberrecker et al., 1967).

Oxygen (O_2) acting upon the ZnO surface, after it has been heated in vacuum or in hydrogen, decreases the conductance of this material even at room temperature (Heiland et al., 1959).

The conductance of the surface layer may also be changed due to the diffusion of lattice defects. Heiland (1982) attributes this effect to the formation of additional oxygen vacancies (V_O^{+2}) in the ZnO, after oxygen ions are freed from the crystal lattice to the atmosphere according to

$$O_{(lattice)}^{-2} \rightarrow V_O^{+2} + 2e^- + \tfrac{1}{2}O_2(gas). \tag{6.4}$$

The significance of these effects increases with temperature (above 870 K), since then they proceed at a high rate and all the additives contained in the ZnO and disturbing the sensor readings have already been ionized. On the other hand, the structure of the surface layer of the crystal may undergo irreversible changes in the given atmosphere. For example, in SnO_2 heated at a temperature of 570 K under a hydrogen pressure of 0.1 Pa (in vacuum), aggregates of metallic tin form down to a depth of several nanometres. The effects occurring during the activation operation that consist of 'sensitizing' the sensor before it is set to operation, such as the heating of the ZnO-based oxygen sensor in hydrogen mentioned earlier, include:

— establishing a high donor concentration in the near-surface layer; during the operation of the sensor, the donors enhance the changes of its conductance;

— reducing the concentration of trap levels which decrease the sensor conductance;

— introducing certain structural changes that enhance the adsorption.

Göpel et al. (1988) and Schierbaum et al. (1988, 1991a,b) studied how an addition of a metal and oxygen affects the structure of the near-surface region in SnO_2 and TiO_2. They used various spectroscopic techniques and found that even noble metals such as Pt and Pd diffuse into oxide in the form of Pt^{+2} or Pt^{+4} ions. This results in the Schottky barrier being decreased. The broadening of the Pd–SnO_2 interface can even be detected after annealing at a temperature as low as 670 K. Hence, it was concluded that the interface structure may undergo changes under normal operating conditions (typically 370–670 K in a reducing atmosphere) of the sensor. The bulk structure of the oxide may also be changed, as has been shown when measuring the J–V characteristics of Pt/TiO_2 junctions formed in TiO_2. Göpel et al. and Schierbaum et al. also measured the variations of the conductance, work function and catalytic activity. All the phenomena mentioned above can affect the sensitivity, selectivity and stability of the sensor.

Mizsei and Harsanayi (1983) describe how the work function of the electrons contained in thin SnO_2 layers deposited by sputtering varies as a function of the composition of the atmosphere. As a result of the adsorption of gases, the work function of the electrons at the surface of the n-type SnO_2 semiconductor changes so that the concentration of free carriers change and a potential barrier forms.

Brailsford and Logothetis (1985) have proposed a sophisticated mathematical model of the behaviour of a porous ceramic gas TiO_2-based sensor. They have shown that the effect of the catalyst molecules may be controlled based on the classical theory of diffusion in porous materials. TiO_2-based sensors are used for measuring the air/fuel ratio in combustion engines. They exhibit a stepwise resistance change when the p_{CO}/p_{O_2} ratio reaches a certain definite value. How much their resistance then changes depends on the presence of a catalyst (Pt or Rh) and the temperature at which it is measured (Fig. 6.21). Moreover, in the presence of a catalyst, the resistance jump (at lower temperatures) shifts towards greater values of the p_{CO}/p_{O_2} ratio.

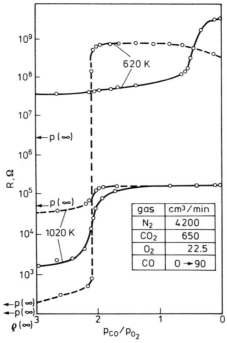

Fig. 6.21. Resistance of a TiO_2-based sensor (solid line — TiO_2 without catalytic inclusions, dashed line — TiO_2 containing Pt–Rh catalyst) versus CO concentration in the reference gas (the gas composition specified in the figure), as affected by the presence of a catalyst at various temperatures (reprinted from Brailsford and Logothetis, 1985, by permission of Elsevier Sequoia S.A.)

When constructing their model, Brailsford and Logothetis have taken into account present knowledge about adsorption and the effect of sample porosity upon the properties of the sensor. They explained the jump in the sensor resistance in terms of the deviations from stoichiometry and the changes that occur in the defect structure of TiO_2. Taking into account the adsorption reaction and using the defect equilibrium equations derived for a given composition of the atmosphere, they calculated the resistance value at which its jump begins. The presence of a catalyst affects the equilibrium of the reactions. Using numerical calculations, Brailsford and Logothetis have obtained curves similar to those shown in Fig. 6.21. Clifford and Tuma (1982/83a,b) have constructed a similar model describing the Figaro SnO_2-based sensors.

In view of the high complexity of the adsorption and catalysis processes and the difficulties in defining precisely the sensor material (i.e., describing the structure of its surface, the deviations from stoichiometry of its composition, etc.), we must realize the difficulties in constructing a complete mathematical model that will explain the behaviour of gas sensors. The urgent need for these sensors is, however, a strong stimulus for further theoretical studies in this field.

6.3 SELECTIVITY

The selectivity of gas sensors, i.e., their ability to react to the presence or to a given concentration of one or more selected gases, is one of their most important parameters and, at the same time, presents one of the greatest difficulties to overcome in their manufacture. The selectivity is so important since, when it is too high, the sensor may raise a spurious alarm, or, when too poor, the sensor may not react to a real danger. Fortunately, in most cases, the sensor is used to respond to a given gas in situations where the other gases to which it is sensitive are absent in the atmosphere.

Morrison (1987) describes four common techniques that permit the fabrication of sensors of high selectivities. These are:
— controlling the temperature,
— using surface additives,
— using selective filters,
— using appropriate catalysts and their activators.

SnO_2-based sensors usually operate at elevated temperatures produced, for example, by a heater installed in the Al_2O_3 tube. The kinetics of reaction (6.3) obviously depends on the nature of the RH_2 reagent. Generally, it can be assumed that each such reagent implies a different temperature at which the sensitivity of the sensor is greatest. If the temperature is too low, reaction (6.3) runs slowly; if it is too high, the oxidizing reaction proceeds very quickly and the RH_2 concentration at the surface is controlled by diffusion. If this is so, from the point of view of the sensor this concentration tends to zero leading to low sensitivity. Yamazoe et al. (1983) have determined the maximum sensitivity temperatures for sensors fabricated of various materials (SnO_2 doped with a metal) for several gases. Their results are shown in Table 6.2. The measure of the sensitivity of SnO_2 sensors is the maximum ratio of their resistance in air to their resistance in the gas examined. The table gives the values of this ratio and, in brackets, the temperature (in °C) at which they were measured.

Table 6.2 Sensitivities of SnO_2-based sensor doped with various metals (according to Yamazoe et al., 1983)

Additive	0.02% CO	0.8% H_2O	0.2% C_3H_8	0.5% CH_4
None	4 (200)	37 (200)	49 (350)	20 (450)
Mn (2.0%)	1	3 (300)	3 (300)	2 (450)
Co (2.0%)	1	3 (300)	2 (300)	2 (400)
Ni (2.0%)	7 (150)	169 (250)	67 (300)	9 (350)
Cu (0.5%)	7 (150)	98 (300)	48 (325)	20 (350)
Ru (0.5%)	2 (150)	63 (150)	67 (150)	3 (300)
Rh (2.0%)	1	11 (150)	4 (200)	3 (300)
Pd (0.5%)	12 (RT)	119 (150)	75 (250)	20 (325)
Ag (0.5%)	8 (100)	666 (100)	89 (350)	24 (300)
Pt (0.5%)	136 (RT)	3600 (RT)	38 (275)	19 (300)
$La_{0.6}Sr_{0.4}$·CoO_3 (2.0%)	34 (100)	184 (250)	71 (250)	9 (350)

RT — room temperature.

By adjusting the temperature we may use the same sensor for detecting different gases (Lalauze et al., 1983, 1984). Since, however, the temperature ranges within which the sensor sensitivity is high are wide, this method is not often employed. Torvela et al. (1991a) have recently shown that the presence of NO and SO_2 can disturb the performance (CO detection) of tin oxide gas sensors at low temperatures. When the temperature of the sensor exceeds 770 K, the interfering effect of these gases becomes negligible.

Metal additives also affect the stability of gas sensors. Having examined 31 metallic dopants, Matsuura and Takahata (1991) found that the stability of the Figaro sensor can be improved by adding rhenium and vanadium.

In his review papers, Morrison (1986, 1987) discusses some of the methods described above for increasing the selectivity of the ceramic gas sensors.

In certain cases, it is advantageous to cover the surfaces of the sensor grains with a substance that absorbs the gas to be detected or reacts with it.

This substance may, for example, be a hygroscopic salt in humidity sensors or sulphanilic acid in NO_2 detecting sensors. The electrons contained in this acid pass into the semiconductor. For this reasons, the NO_2 sensors are fabricated of a p-type semiconductor, such as NiO. As a result of the reaction between the surface of the sensor and NO_2, the electrons return to the gas and the conductance of the sensor increases (Morrison, 1982).

Another method uses selective filters which only permit the substance to be detected to reach the sensor. These filters are made of the following substances:

— silica in hydrogen-detecting sensors (hydrogen can flow through SiO_2 much more easily than hydrocarbons), cf. Fukui and Komatsu (1983);

— tin oxide in CH_4-detecting sensors (Komori et al., 1983);

— metallized membranes through which CO_2 passes more freely than oxygen does (Nagashima and Suzuki, 1984; Hassan and Tadros, 1985);

— ZrO_2, which at elevated temperatures may be used to pass oxygen alone (Okamoto et al., 1980);

— zeolites, which adsorb many gases preventing them from reaching the sensor (Advani and Jordan, 1980; Allman and Khilnani, 1983);

— Pt or Pd films which pass hydrogen.

A method widely used for increasing the selectivity of gas sensors involves introducing catalysts into the sintered structure. The mechanism of catalysis is not yet well known. In general terms, it has been described in Section 2.4. The catalyst, whose small grains cover the surfaces of the grains of the ceramic material, facilitates reaction (6.3). This is illustrated in Fig. 6.22a, from which we can easily see that the products of the oxidizing reaction may escape into the atmosphere so that the sensor does not detect them. This effect poses a serious problem, since it may even decrease the sensor sensitivity. Morrison (1987) mentions two ways in which the catalyst may affect the properties of the grain surface and of the intergranular contacts. One is 'spillover' and the other is Fermi level control.

Spillover is a well known effect in heterogeneous catalysis, which probably occurs most often when metallic catalysts, such as Pt and Pd, are used. The reaction products move over the catalyst surface towards the substrate, i.e., towards the surface of the ceramic grains (Fig. 6.22b). This effect has been examined for hydrogen ions (Sermon and Bond, 1973) and oxygen. The reactions occurring on the catalyst and initiating more complicated reactions are

$$2Pt \cdot + H_2 \rightarrow 2Pt : H \qquad (6.5)$$

$$2Pt \cdot + O_2 \rightarrow 2Pt{-}O \qquad (6.6)$$

These reactions involve the dissociation of hydrogen and oxygen with the

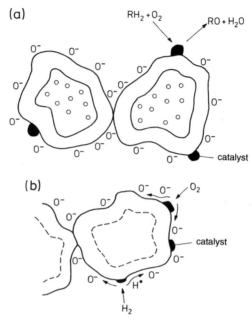

Fig. 6.22: (a) In certain reactions, the catalyst does not affect the surface conductance of sensor grains; (b) due to the spillover effect, electric charge is delivered onto the grain surface whereby changing the surface conductance (reprinted from Morrison, 1987, by permission of Elsevier Sequoia S.A.)

formation of covalent (6.6) or ionic (6.5) bonds. The oxygen and hydrogen then migrate onto the grain surface of the ceramic, where further reactions occur

$$O + e^- \rightarrow O^- \tag{6.7}$$

$$2H \cdot + O^- \rightarrow H_2O + e^- \tag{6.8}$$

$$RH_2 + 2O^- \rightarrow RO + H_2O + 2e^- \tag{6.9}$$

These reactions affect the sensor resistance. In order that the catalyst may act effectively, its grains should be small and uniformly distributed over the ceramic grain surfaces.

Some examples of the spillover phenomena occurring on oxides with metallic clusters are given by Kohl (1990).

The other way in which the catalyst can improve the sensor behaviour is through *control of the Fermi level*. In general terms, this involves the adsorption of oxygen on the catalyst which removes electrons from it; in turn, the catalyst removes electrons from the semiconducting ceramic grain (Fig. 6.23). When a catalyst grain is deposited on the surface of the ceramic grain, surface states form and the catalyst controls the height of the

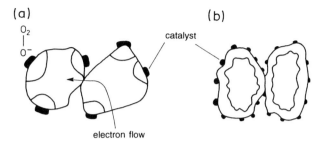

Fig. 6.23. Catalyst dispersion over the grain surface: (a) insufficient dispersion of the catalyst, (b) the whole grain surface covered with the catalyst (reprinted from Morrison, 1987, by permission of Elsevier Sequoia S.A.)

potential barrier present in this region. The oxygen ions adsorbed directly on the ceramic grain are neglected here. In order that the catalyst may act effectively, it should be dispersed over the ceramic grain surface as widely as possible so that the potential barrier covers it entirely. This situation is shown in Fig. 6.23b. The depletion layer usually extends down to a depth of 50 to 200 nm in the semiconductor body. The spacing between the individual catalyst particles should not be greater than 50 nm.

Examples of Fermi level control have been given by Kimoto and Morrison (1977). They examined the behaviour of molybdenum trioxide as a catalyst, with TiO_2 as the substrate. They have determined how the properties of the sensor depend on the degree of dispersion of the catalyst.

When noble metals are used as the catalysts of chemical reactions, we may assume that each surface atom of the metal is active in the catalysis. With oxide catalysts, the best performance is obtained with p-type semiconductors that contain weakly bonded lattice oxygen ions. The oxides used for fabricating gas sensors, such as SnO_2, TiO_2 and ZnO, in which the surface oxygen ions are strongly bonded, show much worse behaviour as catalysts.

As mentioned earlier, oxygen can be adsorbed in various forms: as O_2, O_2^-, O^- and O^{-2}. The oxygen placed in the lattice sites (O^{-2}) of an oxide catalyst can also be reactive with reducing gases, whereas O_2 molecules, which are not dissociated, and adsorbed O^{-2}, which appear in small concentrations, are usually inactive. The presence of O_2^- and O^- ions can be detected using electron spin resonance (ESR). Each ion gives a characteristic signal. In this way it has been found that when a reducing agent is placed where these ions are present, O^- ions disappear very quickly as compared with O_2^- ions, which means that the former are much more active.

Analysis of the catalysis that proceeds in ceramic sensor materials differs from that applicable to catalytic chemical reactions. When dealing with ceramics, our aim is to find appropriate reagents and to prevent reactions with other agents. We are, thus, interested in the first stages of the reactions. When analysing catalytic chemical reactions, we aim at producing given final products, so that the whole course of the reaction should be controlled.

6.4 OTHER MATERIALS SUITABLE FOR FABRICATING GAS SENSORS

Work on fabricating good performance gas sensors is carried out very extensively at the present. In view of the difficulties in producing sensors of high selectivities and stabilities, many new materials are thoroughly examined. Below, we shall present a few examples of research work carried out in this field.

Gas sensors intended for certain applications appear to best fabricated by the thin-layer techniques. Mokwa et al. (1985), for example, describe a sensor in which the active material is a 40 nm thick tin oxide layer, evaporated under vacuum and then oxidized in a controlled manner. The conductance of this layer varies when the sensor is exposed to AsH_3, even in small concentration; AsH_3 is used for the production of epitaxial layers on GaAs crystals and is very poisonous. In the presence of a few ppm of AsH_3 in air at a temperature of 710 K, the conductance of the tin layer increases markedly. Thin SnO_2 layers (deposited by the sputtering technique), on the other hand, change their conductance when exposed to carbon monoxide (Winidischmann and Mark, 1979) and to hydrogen (Dibbern et al., 1986).

Bott et al. (1984) report on the variation of the conductance of ZnO single crystals in response to the presence of CO, CH_4, H_2 and H_2O. ZnO appears to compete against SnO_2 as the material suitable for fabricating high-performance gas sensors and, hence, its properties have been examined by many investigators (Lagowski et al., 1977; Firth et al., 1974, 1975; A. Jones et al., 1984; T. A. Jones et al., 1985; Heiland and Kohl, 1985; Seiyama et al., 1962; Heiland, 1982; Cossement et al., 1986; Pizzini et al., 1989).

Other materials suitable for fabricating gas sensors are TiO_2 (Heiland and Kohl, 1985; Brailsford and Logothetis, 1985; Logothetis and Kaiser, 1983) or the ceramic $TiO_2 + SnO_2$ (Yamamoto and Shimizu, 1982). Kuwabara and Ide (1987) used porous $BaTiO_3$ ceramic for measuring the concentration of CO. Zinc and germanium oxynitrides may be used for measuring the concentration of ammonia (Rosse et al., 1986).

Haruta et al. (1986) report on thick-layer α-Fe_2O_3 sensors, doped with TiO_2 and with gold precipitates as showing a high sensitivity to CO. Lantto et al. (1986) mention WO_3 (which is an n-type semiconductor) as sensor-type material, and Kwang-Soo Yoo and Hyung-Jin Jung (1986) describe the gas sensing properties of In_2O_3 sintered with Al_2O_3, Y_2O_3, SnO_2 and $PdCl_2$. Because of the limited volume of this book, these works cannot be described in detail here. The interested reader is referred to the respective references.

REFERENCES

Advani G. J., and Jordan A. G. (1980), Thin films of SnO_2 as a solid state gas sensors, *J. Electron. Mater.*, **9** (1), 29.

Allman C. E. and Khilnani G. (1983), Poison resistant combustible gas sensors, *Adv. Instrum.*, **38**, 399.

Amamoto T., Tanaka K., Takahata K., Matsuura S. and Seiyama T. (1990), A fire detection experiment in a wooden house by SnO_2 semiconductor gas sensors, *Sensors and Actuators B-Chemical*, **1** (1-6), 226.

Aucouturier J. L. (ed.) (1986), *Proc. 2nd Int. Meeting on Chemical Sensors*, Bordeaux, France.

Bott B., Jones T. A. and Mann B. (1984), The detection and measurement of CO using ZnO single crystals, *Sensors and Actuators*, **5** (1), 65.

Brailsford A. D. and Logothetis E. M. (1985), A steady-state diffusion model for solid-state gas sensors, *Sensors and Actuators*, **7** (1), 39.

Bühling D. (1985), Keramische Sensoren, *Radio-Fernsehen Elektron.*, **34** (8), 484.

Clifford P. K. and Tuma D. T. (1982/83a), Characteristics of semiconductor gas sensors. I. Steady state gas response, *Sensors and Actuators*, **3** (3), 233.

Clifford P. K. and Tuma D. T. (1982/83b), Characteristics of semiconductor gas sensors. II. Transient response to temperature changes, *Sensors and Actuators*, **3** (3), 255.

Coles G. S. V., Gallagher K. J. and Watson J. (1985), Fabrication and preliminary tests on tin(IV) oxide-based gas sensors, *Sensors and Actuators*, **7** (2), 89.

Coles G. S. V., Bond S. E. and Williams G. (1991a), Selectivity studies and oxygen dependence of tin(IV) oxide-based gas sensors, *Sensors and Actuators B-Chemical*, **4** (3–4), 485.

Coles G. S. V., Williams G. and Smith B. (1991b), Selectivity studies on tin oxide-based semiconductor gas sensors, *Sensors and Actuators B-Chemical*, **3** (1), 7.

Cossement D., Pierson E., Streydio J. M., Pirotte D. and Delmon B. (1986), Sprayed ZnO gas sensor, in: Aucouturier J. L. (ed.) (1986), *Proc. 2nd Int. Meeting on Chemical Sensors*, Bordeaux, France, 183.

Dibbern U., Kursten G. and Willich P. (1986), Gas sensitivity, sputter conditions and stoichiometry of pure tin oxide layers, in: Aucouturier J. L. (ed.) (1986), *Proc. 2nd Int. Meeting on Chemical Sensors*, Bordeaux, France, 127.

Firth J. G., Jones A. and Jones T. A. (1974), Solid state sensors for gas detection, *Proc. IERE Conf. Env. Sensors and Applications*, London, 57.

Firth J. G., Jones A. and Jones T. A. (1975), Solid state sensors for carbon monoxide, *Ann. Occ. Hyg.*, **18**, 63.

Fukui K. and Komatsu K. (1983), H_2 gas sensor of sintered SnO_2, in: Seiyama T., Fueki K., Shiokawa J. and Suzuki S. (eds.) (1983), *Chemical Sensors, Anal. Chem. Symposia Series*, Vol. 17, Elsevier, Amsterdam, 52.

Göpel W., Wiemhofer H. D., Kirner U. and Rocker G. (1988), Surface and bulk properties of TiO_2 in relation to sensor applications, *Solid State Ionics*, **28–30**, 1423.

Haberrecker K., Mollwo E., Schreiber H., Hoinkes H., Nahr H., Lindner P. and Wilsch H. (1967), The ZnO-crystal as sensitive and selective detector for atomic hydrogen beams, *Nucl. Instr. and Meth.*, **57** (1), 22.

Haruta M., Kobayashi T., Sano H. and Nakane M. (1986), A novel CO sensing semiconductor with coprecipitated ultrafine particles of gold, in: Aucouturier J. L. (ed.) (1986), *Proc. 2nd Int. Meeting on Chemical Sensors*, Bordeaux, France, 179.

Hassan S. S. M. and Tadros F. S. (1985), Performance characteristics and some applications of the nitrogen oxide gas sensor, *Analyt. Chem.*, **57** (1), 162.

Heiland G. (1957), Zum Einfluss von Wasserstoff auf die elektrische Leitfähigkeit an der Oberfläche von ZnO-Kristallen, *Z. Physik*, **148**, 15.

Heiland G., Mollwo E. and Stockmann F. (1959), Electronic processes in ZnO, in: Seitz F. and Turnbull D. (eds.) (1959), *Solid State Physics*, Vol. 8, Academic Press, New York, 191.

Heiland G. (1982), Homogenous semiconducting gas sensors, *Sensors and Actuators*, **2** (4), 343.

Heiland G. and Kohl D. (1985), Problems and possibilities of oxidic and organic semiconductor gas sensors, *Sensors and Actuators*, **8** (3), 227.

Ichinose N. (1985), Electronic ceramics for sensors, *Am. Ceram. Soc. Bull.*, **64** (12), 1581.

Ichinose N. (1986), Ceramic materials herald the future of sensors, *J. Electr. Eng.*, **23** (230), 93.

Ichinose N. (1987), Ceramic sensors in the microprocessor industry, in: Vincenzini P. (ed.) (1987), *High-Tech Ceramics*, Elsevier, Amsterdam, 95.

Jones A., Jones T. A., Mann B. and Firth J. G. (1984), The effect of the physical form of the oxide on the conductivity changes produced by CH_4, CO and H_2O on ZnO, *Sensors and Actuators*, **5** (1), 75.

Jones T. A., Firth J. G. and Mann B. (1985), The effect of oxygen on the electrical conductivity of some metal oxides in inert and reducing atmospheres at high temperature, *Sensors and Actuators*, **8** (4), 281.

Kimoto K. and Morrison S. R. (1977), Electron and oxygen exchange between reactants and bismuth molybdate, *Z. Phys. Chem. N. F.*, **108** (1), 11.

Kohl D. (1990), The role of noble-metals in the chemistry of solid-state gas sensors, *Sensors and Actuators B-Chemical*, **1** (1-6), 158.

Komori N., Sakai S. and Komatsu K. (1983), Sintered SnO_2 sensor for methane, in: Seiyama T., Fueki K., Shiokawa J. and Suzuki S. (eds.) (1983), *Chemical Sensors, Anal. Chem. Symposia Series*, Vol. 17, Elsevier, Amsterdam, 47.

Kulwicki B. M. (1984), Ceramic sensors and transducers, *J. Phys. Chem. Solids*, **45** (10), 1015.

Kuwabara M. and Ide T. (1987), CO gas sensitivity in porous semiconducting barium titanate ceramics, *Am. Ceram. Soc. Bull.*, **66** (9), 1401.

Kwang-Soo Yoo and Hyung-Jin Jung (1986), Gas sensing characteristics of semiconducting materials based on In_2O_3 depending on composition changes, in: Aucouturier J. L. (ed.) (1986), *Proc. 2nd Int. Meeting on Chemical Sensors*, Bordeaux, France, 230.

Lagowski J., Sproles E. S. and Gatos H. S. (1977), Quantitative study of the charge transfer in chemisorption. Oxygen chemisorption on ZnO, *J. Appl. Phys.*, **48** (8), 3566.

Lalauze R., Bui N. D. and Pijolat C. (1983), Chemical gaseous treatment of SnO_2 for a selective detection of gas, in: Seiyama T., Fueki K., Shiokawa J. and Suzuki S. (eds.) (1983), *Chemical Sensors, Anal. Chem. Symposia Series*, Vol. 17, Elsevier, Amsterdam, 47.

Lalauze R. and Pijolat C. (1984), A new approach to selective detection of gas by an SnO_2 solid-state sensor, *Sensors and Actuators*, **5** (1), 55.

References

Lantto V. E., Romppainen P. S. and Leppavuori S. T. (1986), Investigative study on effect of pure and mixed gas systems on resistive type metal oxide sensors under different experimental conditions, in: Aucouturier J. L. (ed.) (1986),*Proc. 2nd Int. Meeting on Chemical Sensors*, Bordeaux, France, 186.

Logothetis E. M. and Kaiser W. J. (1983), TiO_2 film oxygen sensors made by chemical vapour deposition from organometallics, *Sensors and Actuators*, **4** (3), 333.

Matsushima S., Maekawa T., Tamaki J., Miura N. and Yamazoe N. (1992), New methods for supporting palladium on a tin oxide gas sensor, *Sensors and Actuators B-Chemical*, **9** (1), 71.

Matsuura Y. and Takahata K. (1991), Stabilization of SnO_2 sintered gas sensors, *Sensors and Actuators B-Chemical*, **5** (1–4), 205.

McAleer J. F. (1985), Tin oxide gas sensors: Use of Seebeck effect, *Sensors and Actuators*, **8** (3), 251.

McAleer J. F., Moseley P. T., Norris J. O. W., Scott G. V., Tappin G. and Bourke P. (1986), A Seebeck effect gas sensor, in: Aucouturier J. L. (ed.) (1986), *Proc. 2nd Int. Meeting on Chemical Sensors*, Bordeaux, France, 201.

Mizsei J., Harsanayi J. (1983), Resistivity and work function measurements on Pd-doped SnO_2 sensor surface, *Sensors and Actuators*, **4** (3), 397.

Mokwa W., Kohl D. and Heiland G. (1985), An SnO_2 thin film for sensing arsine, *Sensors and Actuators*, **8** (2), 101.

Morrison R. S. (1982), Semiconductor gas sensors, *Sensors and Actuators*, **2** (4), 329.

Morrison R. S. (1986), Selectivity in semiconductor sensors, in: Aucouturier J. L. (ed.) (1986), *Proc. 2nd Int. Meeting on Chemical Sensors*, Bordeaux, France, 39.

Morrison R. S. (1987), Selectivity in semiconductor gas sensor, *Sensors and Actuators*, **12** (4), 425.

Moseley P. T. (1992), Materials selection for semiconductor gas sensors, *Sensors and Actuators B-Chemical*, **6** (1), 149.

Nagashima K. and Suzuki S. (1984), Solid state electrochemical detector for carbon monoxide at sub-ppm concentrations, *Anal. Chim. Acta*, **162**, 153.

Okamoto H., Obayashi H. and Kudo T. (1980), Carbon monoxide gas sensor made of stabilized zirconia, *Solid State Ionics*, **1** (3/4), 319.

Oyabu T., Ohta Y. and Kurobe T. (1986), Tin oxide gas sensor and countermeasure system against accidental gas leaks, *Sensors and Actuators*, **9** (4), 301.

Paria M. K. and Maiti H. S. (1982), Electrical conductivity and defect structure of polycrystalline tin oxide doped with antimony oxide, *J. Mat. Sci*, **17** (11), 3275.

Pizzini S., Palladino M., Butta N. and Narducci D. (1989), Thick-film ZnO resistive gas sensors—Analysis of their kinetic-behavior, *J. Electrochem. Soc.*, **136** (7), 1945.

Romppainen P., Torvela H., Vaananen J. and Leppavuori S. (1985), Effect of CH_4, SO_2 and NO on the CO response of an SnO_2-based thick film gas sensor in combustion gases, *Sensors and Actuators*, **8** (4), 271.

Rosse G., Ghers M., Guyader J., Laurent Y. and Colin Y. (1986), Selective detection of ammonia by semiconducting pellets of zinc and germanium oxynitrides, in: Aucouturier J. L. (ed.) (1986), *Proc. 2nd Int. Meeting on Chemical Sensors*, Bordeaux, France, 134.

Saaman A. A. and Bergveld P. (1985), A classification of chemically sensitive semiconductor devices, *Sensors and Actuators*, **7** (2), 75.

Schierbaum K. D., Wiemhofer H. D. and Göpel W. (1988), Defect structure and sensing mechanism of SnO_2 gas sensors—Comparative electrical and spectroscopic studies, *Solid State Ionics*, **28–30**, 1631.

Schierbaum K. D., Weimar U., Göpel W. and Kowalkowski R. (1991a), Conductance, work function and catalytic activity of SnO_2-based gas sensors, *Sensors and Actuators B-Chemical*, **3** (3), 205.

Schierbaum K. D., Kirner U. K., Geiger J. F. and Göpel W. (1991b), Schottky-barrier and conductivity gas sensors based upon Pd/SnO_2 and Pt/TiO_2, *Sensors and Actuators B-Chemical*, **4** (1–2), 87.

Seiyama T., Kato A., Fujishi K. and Nagatani M. (1962), A new detector for gaseous components using semiconducting thin films, *Analyt. Chem.*, **34** (11), 1502.

Seiyama T., Fueki K., Shiokawa J. and Suzuki S. (eds.) (1983), *Chemical Sensors, Anal. Chem. Symposia Series*, Vol. 17, Elsevier, Amsterdam, 52.

Sermon P. A. and Bond G. C. (1973), Hydrogen spillover, *Catalysis Rev.*, **8** (2), 211.

Taguchi N. (1972), Gas detection device, British Patent 1,280,809.

Torvela H., Leppavuori S. and Harkoma A. (1989), Detection of the concentration of CO using SnO_2 gas sensors in combustion gases of different fuels, *Sensors and Actuators*, **17** (3–4), 369.

Torvela H., Harkomamattila A. and Leppavuori S. (1990), Characterization of a combustion process using catalyzed tin oxide gas sensors to detect CO from emission gases, *Sensors and Actuators B-Chemical*, **1** (1–6), 83.

Torvela H., Huusko J. and Lantto V. (1991a), Reduction of the interference caused by NO and SO_2 in the CO response of Pd-catalyzed SnO_2 combustion gas sensors, *Sensors and Actuators B-Chemical*, **4** (3–4), 479.

Torvela H., Pijolat C. and Lalauze R. (1991b), Dual response of tin oxide gas sensors characteristic of gaseous carbon tetrachloride, *Sensors and Actuators B-Chemical*, **4** (3–4), 445.

Uchino K. (1986), Electrostrictive actuators: materials and applications, *Am. Ceram. Soc. Bull.*, **65** (4), 647.

Uematsu K., Kato Z., Uchida N. and Saito K. (1987), Electrical conductivity of antimony-doped tin dioxide prepared by hot isostatic pressing, *J. Am. Ceram. Soc.*, **70** (7), C-142.

Watson J. (1984), The tin oxide gas sensor and its applications, *Sensors and Actuators*, **5** (1), 29.

Watson J. and Yates R. A. (1985), A solid-state gas sensor, *Electronic Engineering*, **57** (701), 47.

Winidischmann H. and Mark P. (1979), A model for the operation of a thin film SnO_2 conductance modulation carbon monoxide sensor, *J. Electrochem. Soc.*, **126** (4), 627.

Yamamoto T. and Shimizu H. (1982), Some considerations on stability of electrical resistance of the TiO_2/SnO_2 ceramic moisture sensor, *IEEE Trans. on Comp. Hybr. and Manuf. Techn.*, **CHMT-5** (2), 38.

Yamazoe N., Kurokawa Y. and Seiyama T. (1983), Effects of additives on semiconductor gas sensors, *Sensors and Actuators*, **4** (2), 283.

Yannopoulos L. N. (1987), Antimony-doped stannic oxide-based thick-film gas sensors, *Sensors and Actuators*, **12** (1), 77.

Yasunaga S., Sunahara S. and Ihokura K. (1986), Effects of tetraethyl orthosilicate binder on the characteristics of an SnO_2 ceramic-type semiconductor gas sensor, *Sensors and Actuators*, **9** (2), 133.

CHAPTER 7

Overview

The materials described in the present book have one feature in common, namely the fact that their useful properties are determined by the properties of their grain boundaries. The phenomena occurring at grain boundaries are relatively well known and have been described using various models, such as the model of a metal-semiconductor junction. When, however, we deal with a polycrystal, which in addition contains considerable amounts of impurities and, often, has a polyphase structure, the interpretation of these phenomena becomes much more difficult. The examination of the active regions of grain boundaries is also difficult in these crystals, and so is the development of appropriate technology.

The subtle energy structure of the grain boundary is determined by the configuration of atoms, both the atoms of dopants and the atoms of the basic phase, such as oxygen atoms. On the grain surfaces in a ceramic material, an adsorption layer forms.

Investigators studying the properties of these materials face many difficulties, of which the most important is the degradation effect, i.e., a slow variation of their properties during exploitation. The factors responsible for this effect are numerous. In the present author's opinion, an important role is played by the oxygen atoms present in the ceramic material. All the materials described in the present book are oxide-based materials. The effect of the oxygen atoms upon the properties of the material depends on their distribution throughout the material structure. How the oxygen atoms placed in the lattice sites affect the stoichiometry of the material has been discussed in the book. The adsorption of oxygen on the surface of the ceramic material and its practical application in fabricating gas sensors have also been described. Oxygen atoms are able to move within the material, especially along the grain boundaries, even at a relatively low temperature. Their movement may be responsible for degradation of all the ceramic materials described here.

When dealing with materials intended for electronics, where miniaturization and high performance are the major requirements, the size of the components approaches the possible limits, as is the case with transistors in an integrated circuit. These limits result from the 'mechanical' nature of their manufacturing process, which consists of depositing layers of differing properties. Basically, the structure of these circuits is two-dimensional. The progress achieved to date is associated with the continuous improvement of the equipment and with the use of new materials, such as GaAs which replaces silicon. It, however, seems that an ideal structure should be three-dimensional, and that its properties should vary within regions whose dimensions are close to a few interatomic spacings. This would permit a revolutionary advance to be made in both the miniaturization of integrated circuits and the speed of their operation.

In order to fabricate such a structure, we must be able to control the properties of microscopic regions within a monolithic material. The first step towards this technology may be the technique of fabrication of ceramic materials in which intermediate layers are formed at the grain boundaries of the basic phase. Still new findings made in the field of ceramics permit us to expect that new promising properties and features will be discovered. The recent example is the discovery of high-temperature superconduction in perovskite-type ceramic materials. This discovery, considered to be epoch-making in the history of science, raised an avalanche of interest in ceramic materials throughout the world. It then took only a year to discover materials that exhibit the superconducting state above the temperature of liquid nitrogen (77 K).

It is difficult to imagine the revolution that would take place in engineering when superconducting ceramic materials are fabricated on a large scale. They will certainly find application in electronics, perhaps in the fabrication of GBBL capacitors with superconducting grain bulks, or varistors that turn into the superconducting state above a certain voltage, or else thermistors in which the superconduction abruptly vanishes above a specified temperature. The art of controlling the properties of microregions in monolithic materials certainly continue to develop. I believe that ceramic materials will become the crucial materials in the technologies of the near future.

Author Index

Abdullah Al K., 85
Advani G.J., 183
Al-Allak H.M., 117
Aleksandrov V.V., 47
Alim M.A., 85
Alles A.B., 50, 76, 139
Allman C.E., 183
Amamoto T., 160
Asokan T., 51
Avdeenko B.K., 47
Avella F.J., 64, 88
Azaroff L.V., 23, 52

Baer W.S., 113, 114
Basu R.N., 125
Bäther K.H., 94
Baumgartner I., 76, 77
Berglund C.N., 113, 114
Bergveld P., 159
Bernasconi J., 63, 64, 82, 83
Bhushan B., 50, 58
Binešti D., 91, 96
Blakely J.M., 31
Blatter G., 79
Boilot J.P., 114
Bonasewicz P., 34, 35, 36, 52
Bond G.C., 183
Bond W.D., 51
Bott B., 186
Bowen L.J., 64, 88
Brailsford A.D., 180, 181, 186
Brauer H., 131, 151
Brook R.J., 118

Brophy J.J., 23
Brückner W., 55, 64, 88, 89
Brugman J.C., 81
Bühling D., 159
Burdick V.L., 50, 76
Butler C.E., 149

Carlson W.G., 59, 94, 95, 96
Casella R.C., 113
Castleberry D.E., 47
Chaput F., 114
Chiang Y.-M., 32, 67, 89, 90, 114, 139, 140
Chin T.S., 123
Chon H., 37
Chul Hyun Yo, 52
Cimino A., 37
Clarke D.R., 70, 71
Clifford P.K., 164, 181
Coles G.S.V., 168, 169, 170, 171
Coolican J., 51
Cooper R.A., 51
Cordaro J.F., 85
Cossement D., 186

Daniels J., 110, 114, 132, 133, 134, 135, 148, 149, 155
Dereń J., 12, 53
Desu S.B., 114, 140
Dibbern U., 186
Dorlanne O., 64
Dorn R., 36
Driear J.M., 51

Author Index

Drofenik M., 116
Dunlop G.L., 64

Eberspachel O., 135
Eda K., 58, 65, 80, 86, 88, 89, 91, 92, 93, 97
Eibl O., 117
Einzinger R., 55, 59, 63, 64, 76, 77, 85, 91, 93
Ejinthoven R.K., 63, 64, 77
Emtage P.R., 56, 64, 65, 83
Eun Hong Lee, 120

Firth J.G., 186
Freer R., 51
Fu S.L., 139
Fujitsu S., 86
Fukui K., 183

Gambino J.P., 71
Gerards A.G., 53, 55
Gerthsen P., 113, 121, 122, 123
Gibbs J.W., 30
Glaister R.M., 150
Glemza R., 36
Goodman G., 148
Göpel W., 34, 35, 36, 179
Graciet M., 47, 69
Grunze M., 36
Gupta T.K., 48, 59, 62, 83, 88, 94, 95, 96

Haanstra H.B., 117, 121
Haayman P.W., 109
Haber J., 30
Haberrecker K., 178
Hagemark K.I., 52, 85
Hampshire S., 51
Hanke L., 136, 137
Hannay N.B., 53, 65
Hardnen J.D., 47
Hardtl K.H., 110, 121, 148
Harmer M.P., 153
Harrison W.A., 23
Harsanayi J., 180
Haruta M., 186

Harwig H.A., 53, 55
Hasegawa A., 139
Hassan S.S.M., 183
Hayashi M., 89, 93
Heiland G., 36, 178, 179, 186
Hench L.L., 19
Hennings D.F.K., 51, 116
Heywang W., 111, 112, 123, 127, 128, 135, 136, 137, 138, 140, 141, 142, 143
Hieda S., 47
Ho I.C., 110, 139
Hoffmann B., 77, 121, 122, 123, 131
Hohenberger G., 78
Honnart F., 54
Hower P.L., 83
Hozer L., 55, 56, 59, 60, 61, 62, 63, 65, 68, 70, 71, 83, 94, 96
Hyung-Jin Jung, 187

Ichinose N., 159, 160
Ide T., 186
Iga A., 71, 88, 89
Igarashi H., 157
Ihrig H., 116, 117, 121, 124, 125, 131
Ikeda J.A.S., 32
Inaba M., 111
Inada M., 71, 74, 75
Inoue H., 126
Ivers-Tiffee E., 51
Iwahara H., 54

Jae Shi Choi, 52
Janitzki A.S., 116
Jones A., 186
Jones T.A., 186
Jonker G.H., 117, 128, 129, 130, 139
Jordan A.G., 183

Kaiser W.J., 186
Keller S.P., 113
Kemenade, van J.T.C., 63, 64, 77
Khilnani G., 183
Ki Hyun Yoon, 120
Kim M.S., 84
Kimoto K., 185
Kimrev H.D., 14

Author Index

Kingery W.D., 15, 32, 67, 71
Kiselev V.F., 39
Kittel C., 19, 20
Klein H.P., 65, 79, 82
Klerk J., 152
Klerk M., 121
Kliewer K.L., 31
Knecht B., 65, 82
Kobayashi M., 47
Koehler J.S., 31
Kohl D., 184, 186
Kokes R.J., 36
Kołacz M., 68
Komatsu K., 183
Komori N., 183
Kosman M., 66
Kostič P., 51
Kröger F.A., 52
Krylov O.V., 39
Kulwicki B.M., 109, 131, 159
Kumamoto K., 123
Kunstmann P., 36
Kuschke R., 151
Kutty T.R.N., 68, 86
Kuwabara M., 115, 118, 119, 121, 122, 123, 124, 125, 126, 127, 140, 141, 142, 143, 157, 186
Kwan T., 37
Kwang-Soo Yoo, 187

Lagowski J., 35, 36, 186
Lalauze R., 182
Lambeck P.V., 139
Lampe U., 34, 35
Langmuir I., 33, 34
Lantto V.E., 186
Lauf R.J., 51
Lee J.J., 88
Lehovec K., 31
Levin E.M., 53, 66
Levine D., 70
Levinson L.M., 46, 47, 56, 62, 65, 69, 79, 97
Li P.W., 52
Lifshitz I.M., 31
Lin T.F., 116
Littbarski R., 34, 35, 36
Logothetis E.M., 180, 181, 186
Lothe H., 36

Lou C.H., 123
Lou L.F., 59, 63, 64

Macdonald J.R., 85, 123
Maeda T., 85
Mahan G.D., 52, 63, 64, 83, 84, 85
Maiti H.S., 125, 171
Many A., 16, 20, 157
Mark P., 186
Masuyama T., 59
Matsui T., 151
Matsuo Y., 117
Matsuoka M., 46, 50, 51, 56, 57, 63, 66, 68, 79, 89
Matsushima S., 167
Matsuura M., 59
Matsuura Y., 182
McAleer J.F., 174, 175
McCutcheon C., 116
Medernach J.W., 53
Meyer A., 117
Milosevič O., 55
Miyayama M., 54
Miyoshi T., 59
Mizsei J., 180
Modine F.A., 65
Mohanty G.P., 52
Mokwa W., 186
Moldenhauer W., 88, 94
Moll J.M., 27
Morris W.G., 59, 63, 66, 67, 70, 71, 85
Morrison R.S., 37, 176, 177, 178, 181, 182, 183, 184, 185
Moseley P.T., 159
Mostaghaci H., 118
Mukae K., 50, 58, 59, 60, 63, 65, 75, 76
Munir Z.A., 15

Nagashima K., 183
Nasrallah M.M., 139
Nemoto H., 123
Neumann G., 52
Nitayama A., 85
Nowotny J., 31

Oda T., 123

Author Index

Okamoto H., 183
Okuma H., 59
Olsson E., 64
Ościk J., 33
Oyabu T., 165, 166

Pajares J., 37
Pampuch R., 11, 12, 13, 15
Paria M.K., 171
Paulson W.M., 87
Payne D.A., 114, 140
Petcold E.G., 66
Philipp H.R., 46, 47, 56, 59, 62, 65, 69, 79, 97
Phule P.P., 114
Pike G.E., 84
Pizzini S., 186
Poeppel R.B., 31
Pruzhinina V.I., 47
Purdes A.J., 131

Raghu N., 68, 86
Rase D.E., 116
Rehme H., 121
Rhodes W.W., 32
Roginsky S., 33
Rohatgi A., 85
Romppainen P., 171, 172
Rosse G., 186
Roth R.S., 53, 66
Roup R.R., 149
Roy R., 116

Saaman A.A., 159
Saburi O., 110
Sachtler W.M., 29
Safronov G.M., 53, 66
Sakabe Y., 157
Sakshaug E.C., 47
Salmon R., 47, 51, 57
Samsonov G.V., 14, 51
Sanders P.J.H., 152
Santen, van R.A., 29
Santhanam A.T., 70
Sasaki H., 117
Sato K., 87, 89, 94

Schierbaum K.D., 179
Schmelz H., 117, 135, 137
Schwing U., 65, 77
Seager C.H., 23, 84
Seitz K., 51, 88
Seiyama T., 186
Selim F.A., 85
Sermon P.A., 183
Shimizu H., 186
Shirley C.G., 87
Shockley W., 16, 20
Shohata N., 85, 89
Simkovich G., 29
Simmons J.G., 23
Snow G.S., 47, 51
Snyder R.L., 53
Sonder E., 97
Stanisič G., 66
Stenton N., 153
Stevenson D.A., 52
Stone F., 33
Strassler S., 79
Stucki F., 86
Suit Das, 14
Sukkar M.H., 52, 86
Sundaram S.K., 149
Suzuki S., 183
Syamaprasad U., 157
Sze S.M., 23
Szymański A., 2, 56, 59, 60, 61, 62, 63, 65, 83

Tadros F.S., 183
Taguchi N., 161
Takada Y., 89
Takagi T., 32, 114, 139, 140
Takahashi T., 54, 90, 91, 96
Takahata K., 182
Takemura T., 51, 59
Tamm I.E., 16, 20, 29
Tanaka J., 85
Tao M., 64
Terasaki Y., 88
Thomma W., 12
Thuillier J.M., 33
Thümmler F., 12
Tomimuro H., 88
Tominaga S., 47

Author Index

Torvela H., 160, 182
Trontelj M., 68
Tseng T.Y., 123
Tsuda K., 63
Tuller H.L., 52, 86
Tuma D.T., 164, 181

Uchino K., 159
Uematsu K., 171

Vandanamme L.K.J., 81
Volkenshtein F.F., 33

Wagner J.B., Jr., 29
Waku S., 151, 156
Wang D.Y., 139
Wang F.F.Y., 8
Wang S.H., 123
Waser R., 157
Watson J., 161, 164, 165
Wernicke R., 110, 148, 150, 152, 153, 154, 155

Wersing W., 140, 141
West J.K., 19
Wheeler R.B., 65
Williams P., 76
Winidischmann H., 186
Winston R.A., 85
Wong J., 55, 63, 66, 68, 69, 70, 74

Yamamoto T., 186
Yamaoka N., 149, 151
Yamaoki H., 59
Yamazoe N., 167, 168, 182
Yan M.F., 30, 31, 32, 37
Yanagida H., 115, 140
Yannopoulos L.N., 174, 175
Yasunaga S., 171, 172, 173, 175, 176
Yates R.A., 161, 163
Yoneda Y., 115
Yoshida J., 89
Young-Sung Yoo, 117, 118

Zeldowitsch Ja., 33
Ziel, van der A., 22, 27

Subject Index

acetamide, 5
activation of the sensor, 179
adsorption, 32–39, 184
    Langmuir theory, 33
    of oxygen
        on $BaTiO_3$, 138
        on sensors, 178
        on ZnO, 33, 34, 96
    physical, 34, 35, 178
AES, 30, 86
applications
    capacitors, 155
    sensors, 159
    thermistors, 111
    varistors, 45
avalanche breakdown, 28

calcination, 7, 50, 119, 120, 174
catalyst, 37
catalytic formation of water, 38
chemisorption, 34, 178
    of oxygen, 140
        on $BaTiO_3$, 139
        on SnO_2, 176–179
        on ZnO, 35–37, 86
conducting paths, 78, 81
co-precipitation, 6, 167, 174, 186
critical doping level, 116, 136, 137
Curie temperature, 112, 113
    control in $BaTiO_3$, 123
Curie–Weiss law, 113, 128
current-voltage characteristics
    thermistors, 120
    varistors, 44, 48, 55
C-V characteristic, 60

dihedral angle, 67, 72, 73
DLTS, 85
doping anomaly in thermistors, 116, 136, 137

EBIC, 77
effective dielectric constant, 154
electrically-active grain boundary, 65, 77, 78, 93, 96
electrodes
    capacitors, 151, 152
    sensors, 161, 162, 167, 168, 171, 173, 174
    thermistors, 119, 121, 138
    varistors, 49, 51
epitaxial layer, 76, 77, 85
ESR, 185

Figaro sensor, 161
freeze drying, 6
furan, 5
furfuryl acid, 5

Gauss law, 21
Gibbs adsorption equation, 30
grain growth, 14
    capacitors, 152

Subject Index

sensors, 176
thermistors, 115, 116
varistors, 68

heterojunction, 19
holes, 83
homojunction, 64, 73, 77, 85, 93

ICTS, 85
image force, 24, 27
impedance spectroscopy, 85, 123, 124, 139, 149
infiltration of second phase in capacitors, 152
ionic conductivity, 54

KCl, 120
Krönig-Penney model, 15

lanthanum oxide, 75, 110, 116, 131–133, 148, 149, 151

manufacturing technique
  capacitors, 150
  sensors, 161–176
  thermistors, 120
  varistors, 50
metal-semiconductor junction, 17
microcapacitor, 149
microstructure model of
  capacitor, 153
  sensor, 178
  varistor, 78
microvaristor, 63, 91
mill
  attrition, 5
  ball, 4, 50, 114
  jet, 5
  vibrating, 5, 6
milling, 4–6, 111, 174

neodymium oxide, 76, 137, 149
nonlinearity coefficient, 44

NTC thermistors, 111, 160

ohmic contact, 18

paraelectric-ferroelectric transformation, 112
phase transformations in varistors, 74
plasma methods, 6
point defects in $BaTiO_3$, 132
Poisson equation, 21, 29, 60
polarization currents, 65
polyvinyl alcohol, 8
potential barrier, 15–20, 26, 36, 37
  capacitors, 148
  sensors, 176, 177, 178, 185
  thermistors, 121, 127, 131, 137–140, 143
  varistors, 63, 85, 86
powder analysis techniques, 7
powder preparation techniques, 4–8
praseodymium oxide, 50, 58, 75, 91, 149
pressing
  axial, 8, 50
  hot isostatic, 51
  isostatic, 9
pyrochlore, 64, 69, 73, 74, 77, 78

recovery of the varistor, 88, 89, 91, 92, 94, 96
Richardson constant, 24, 81
Richardson equation, 23, 24
Richardson plot, 24, 25

samarium oxide, 76, 149
Schottky barrier, *see* potential barrier
Schottky effect, 24, 62
Schottky emission, 24–26, 79
Schottky equation, 24
Schottky plot, 25, 61, 62
Seebeck effect, 37, 174
segregation, 29–32, 86, 140
  driving forces, 31
selective filters, 183
selectivity control, 181
sensor requirements, 160

SHS, 15
SIMS, 30, 97, 139
sintering, 10–15
    capacitors, 148–152
    electron theory, 14
    mass transport, 12
    microwave, 14
    sensors, 160
    stages, 12
    thermistors, 114–119, 134
    varistors, 50, 56, 66, 68, 74
sol-gel method, 6, 51, 114
space charge induced current, 79
spillover, 183, 184
spinel, 64, 69, 70, 74, 77
spray drying, 7
STEM, 32, 67
surface states, 19, 20
    adsorption, 35
    sensors, 176, 184
    Tamm states, 16, 20, 29

thermistors, 128, 131
varistors, 67, 80, 82, 83, 85

tape-casting, 9
thermally-stimulated current, 89, 92, 93, 94
thermal runaway, 87, 88
thermionic emission, 23, 24
    varistors, 79, 80, 81, 82
thiophene, 5
tunnelling of electrons, 26, 27, 80, 82–84

wetting, 62, 72, 73, 152
work function, 17, 24, 179, 180

XPS, 31, 86

Zener breakdown, 27
    varistors, 81